Ensuring Food Safety in the European Union

T0291096

Ensuring Food Safety in the European Union

Marco Silano and Vittorio Silano

CRC Press
Taylor & Francis Group
Boca Raton London New York

CRC Press is an imprint of the
Taylor & Francis Group, an **informa** business

First edition published 2021
by CRC Press
6000 Broken Sound Parkway NW, Suite 300, Boca Raton, FL 33487-2742

and by CRC Press
2 Park Square, Milton Park, Abingdon, Oxon, OX14 4RN

© 2021 Taylor & Francis Group, LLC

CRC Press is an imprint of Taylor & Francis Group, LLC

Reasonable efforts have been made to publish reliable data and information, but the author and publisher cannot assume responsibility for the validity of all materials or the consequences of their use. The authors and publishers have attempted to trace the copyright holders of all material reproduced in this publication and apologize to copyright holders if permission to publish in this form has not been obtained. If any copyright material has not been acknowledged please write and let us know so we may rectify in any future reprint.

Except as permitted under U.S. Copyright Law, no part of this book may be reprinted, reproduced, transmitted, or utilized in any form by any electronic, mechanical, or other means, now known or hereafter invented, including photocopying, microfilming, and recording, or in any information storage or retrieval system, without written permission from the publishers.

For permission to photocopy or use material electronically from this work, access www.copyright.com or contact the Copyright Clearance Center, Inc. (CCC), 222 Rosewood Drive, Danvers, MA 01923, 978-750-8400. For works that are not available on CCC please contact mpkbookspermissions@tandf.co.uk

Trademark notice: Product or corporate names may be trademarks or registered trademarks, and are used only for identification and explanation without intent to infringe.

ISBN: 9780367542818 (hbk)
ISBN: 9780367531201 (pbk)
ISBN: 9781003088493 (ebk)

Typeset in Palantino
by Deanta Global Publishing Services, Chennai, India

CONTENTS

PREFACE

The fundamental right to adequate food, supported by the Committee on Economic, Social and Cultural Rights (UN), is understood as addressing issues of nutrition, safety and cultural acceptability. According to the EU Treaties, food safety is a priority for the European Union that, to this end, collaborates intensively with the Member States and several international organizations such as WHO and FAO.

The present publication deals with the implementation of the food safety approaches announced by the European Union on 12 January 2000 in the "White Paper on Food Safety", with the objective of defining the policies to improve the level of health protection for the consumers of Europe's food.

Since then, during the past 20 years, the European Institutions have adopted and implemented hundreds of legislations covering: (i) all the different aspects of food chain safety from primary agricultural activities to manufacturing, marketing and sale and consumption; (ii) a number of mandatory and voluntary information tools to increase the consumer's ability to make optimal use of foods through a full understanding of their properties; and (iii) the establishment of the European Food Safety Authority with the mandates of scientifically assessing and communicating risks in the food chain to managers, food business operators and consumers, thus reducing levels of uncertainty and disagreement at the European level.

This book aims to present in a structured and clear manner the many initiatives undertaken since then by the European Institutions.

The fact that the publication of this book takes place soon after the adoption and publication of a EU Regulation to further improve the food risk assessment carried out by EFSA, clearly shows that the committal of the European Institutions for achieving the highest level of food safety continues to be very high.

AUTHORS

Dr. Marco Silano graduated in medicine and surgery from the Catholic University of the Sacred Heart, Rome, and the Specialization in General Pediatrics at the University of Milan – III School of Specialization. From May 1999 to March 2000 he was Visiting Fellow at the Division of Nutritional Sciences of Cornell University, Ithaca, New York, where he studied the intestinal metabolism of long-chain polyunsaturated fatty acids on animal and cell models of pediatric diseases. Subsequently, until October 2003, he carried out assistance and research activities first at the Pediatric Clinic of the San Paolo Hospital, Milan, and then at the U.O. of Rheumatology of the IRCCS Children's Hospital Bambino Gesù, Rome. He is currently First Researcher and Director of the Department of Food, Nutrition and Health at the Department of Food Safety, Nutrition and Veterinary Public Health, Istituto Superiore di Sanità Rome, Italy. The main fields of research consist of pediatric nutrition and food intolerances. He is a member of numerous working groups of the Ministry of Health on breastfeeding, infant nutrition and weaning, celiac disease and nutritional security. Since 2008 he has been a member of the Board of the AIC Scientific Committee, of which he was Coordinator in April 2014. M. Silano is the author of more than 70 scientific publications in peer-reviewed international journals, as well as a speaker at more than 100 national and international conferences.

Prof. Dr. Vittorio Silano, after 20 years of activity at the Istituto Superiore di Sanità, Rome (Italy), and an additional similar period of activity at the Italian Ministry of Health as Director General, is currently Contract Professor on "E.U. Food Safety Legislation", at the Medical School, II University of Rome, Italy. Dr. V. Silano has published approximately 500 scientific papers in international journals, focusing on alpha-amylase inhibitors in cereals, cereal proteins toxic in celiac diseases, chemical contaminants, safety of botanicals and botanical preparations, food contact materials, flavourings, food enzymes and public health, and 30 books on food safety scientific and regulatory issues; the most recent, *Novel Foods*, was published in 2018. Dr. V. Silano has also intensively collaborated with the European Food Safety Authority since its inception in 2002 as an external expert and is currently the chairman of its CEP Panel.

1

From the Treaty of Rome in 1957 to the Treaty of Lisbon Enacted in 2009

The European Union Institutions and Legislative Procedures

1.1 INTRODUCTION

As it is stressed in the official website of the European Commission, the European Union (EU) is based on the rule of law. Every action taken by the European Union is founded on treaties that have been approved democratically by all EU Member States. A treaty is a binding agreement between EU Member Countries that sets out EU objectives, rules for institutions, procedures on how decisions are made and the relationship between the EU institutions and their Member Countries. When needed, treaties are amended to make the EU more efficient and transparent, to prepare for new Member Countries and to introduce new areas of cooperation. Under the treaties, EU institutions can adopt legislation, which the Member Countries then implement (EUR-Lex database of EU law).

1.2 FROM THE TREATY OF ROME TO THE TREATY OF LISBON

The "European Economic Community (EEC)" was established with the Treaty of Rome in 1957 with the participation of six countries (Belgium, France, Germany, Italy, Luxembourg and the Netherlands).

About 36 years were needed to move from the EEC with the "Common Market" to the European Community (EC) with the "Single Market" and the four freedoms (i.e. movements of goods, services, people and money) as established by the Treaty of Maastricht in 1993 and the Treaty of Amsterdam in 1999.

About 52 years were necessary to move from the European Community to the current "European Union" as established by the Treaty of Lisbon ratified by all the 28 Member States before entering into force in 2009 (Correra and Silano, 1995; Capelli et al., 2006).

Some difficulties for the further development of the European Union have emerged from the referendum that was held in the UK on 23 June 2016, to decide whether the UK should leave the European Union. As many as 51.9% of the total votes in the UK were in favour of leaving the EU. The referendum turnout was 71.8%, with more than 30 million people voting. The UK has officially left the EU on 31 January 2020 after a general election. An agreement concerning the relationships between the EU and the UK is still under negotiation.

1.3 THE MAIN EU INSTITUTIONS

The main European Union Institutions are the European Commission, the Council and the Parliament.

The European Commission is the executive body of the European Union with the following tasks: (i) to "promote the general interest of the Union", without prejudice to individual Member States; (ii) "to ensure the application of the Treaties" and adopted measures; and (iii) "to execute the budget". The European Commission further holds a virtual monopoly on the legislative initiative, as it proposes all primary EU legislation to the European Parliament (EP) and the Council of the European Union. The Commission is also in charge of proposing and adopting delegated Regulations after consulting expert groups, composed of representatives from each **EU** country, which meet on a regular or occasional basis.

The College of Commissioners is composed of the representatives from all the Member States, and the Commission is structured as General Directorates. The Commissioners are appointed by the Member States. The DG Health is in charge of food safety.

The European Council brings together the Heads of State or Government of the EU Member States in quarterly meetings and seeks to set the overall direction and priorities of the European Union. The Presidents of the European Council and the European Commission are members but have no vote. Although not being a member of the European Council, the High Representative of the Union for Foreign Affairs and Security Policy/Vice-President of the European Commission also takes part in its meetings.

In the EU Council, the government Ministers from each EU Member State meet to discuss, amend and adopt laws and coordinate policies. The Ministers have the authority to commit their governments to the actions agreed upon in these meetings. The Council is the main decision-making institution of the EU and negotiates and adopts primary EU laws, together with the European Parliament, based on the proposals from the European Commission.

The EU Council:

- coordinates EU countries' policies;
- develops the EU's foreign and security policy, based on European Council guidelines;
- concludes agreements between the EU and other countries or international organizations;
- adopts the annual EU budget jointly with the European Parliament.

There are no fixed members of the EU Council as the Council meets in ten different configurations, each corresponding to the policy area being discussed. Depending on the configuration, each country sends its Minister responsible for the policy area under discussion. For example, when the Council meeting on economic and financial affairs (the "Ecofin Council") is held, it is attended by each country's Finance Minister.

The Foreign Affairs Council has a permanent chairperson – the EU High Representative for Foreign Affairs and Security Policy. All other Council meetings are chaired by the relevant minister of the country holding the rotating EU presidency.

Overall consistency is ensured by the General Affairs Council – which is supported by the Permanent Representatives Committee that is composed of EU countries' Permanent Representatives to the EU. All discussions and votes take place in public. To be passed, decisions usually require a qualified majority of 55% of all EU countries representing at least 65% of the total EU population. At least four countries (representing at least 35% of the total EU population) are needed to block a decision. Sensitive topics like foreign policy and taxation require a unanimous vote (all countries pro), whereas a simple majority is required for procedural and administrative issues.

The Parliament, directly elected by the European citizens every four years, is structured as Committees. The number of Members of the European Parliament (MEPs) for each country is roughly proportional to its population, but this is by degressive proportionality: no country can have fewer than 6 or more than 96 MEPs, and the total number cannot exceed 751 (750 plus the President). MEPs are grouped by political affiliation, not by nationality.

The President represents the Parliament to other EU institutions and the outside world and gives the final go-ahead to the EU budget.

Parliament's work comprises two main stages:

- **Committees** – to prepare legislation. The European Parliament has 20 committees and 2 subcommittees, each handling a particular policy area. The committees examine proposals for legislation, and the members of the EP and political groups can put forward amendments or propose to reject a bill. These issues are also debated within the political groups. The Committee on the Environment, Public Health and Food Safety (ENVI Committee) is in charge of food safety.
- **Plenary sessions** – to pass legislation. This is when all the members of the EP gather in the chamber to give a final vote on the proposed legislation and the proposed amendments. These sessions normally are held in Strasbourg, France, for four days a month, but sometimes there are additional sessions in Brussels, Belgium.

The Parliament, together with the Council, adopts the primary Regulations of the EU based on the European Commission proposals, whereas both institutions maintain the right of scrutiny for the delegated Regulations adopted by the European Commission and the Member States.

Two other institutions play important roles in the European Union:

- the Court of Justice of the EU upholds the rule of European law; and
- the Court of Auditors checks the financing of the EU's activities.

In the EU there are a number of other institutions and interinstitutional bodies that play specific roles:

- the European Central Bank is the *central bank* of the European Union countries which have adopted the euro, and it is responsible for European monetary policy;
- the European External Action Service (EEAS) manages the EU's diplomatic relations with other countries outside the bloc and conducts EU foreign and security policy;
- the European Economic and Social Committee represents civil society, employers and employees;
- the European Committee of the Regions represents regional and local authorities;
- the European Investment Bank provides funding for projects that help to achieve the aims of EU, both within and outside the EU, and helps small businesses through the European Investment Fund;
- the European Ombudsman investigates complaints about maladministration by EU institutions and bodies;
- the European Data Protection Supervisor safeguards the privacy of people's personal data;
- the Publications Office publishes information about the EU;
- the European Personnel Selection Office recruits staff for the EU institutions and other bodies;
- the European School of Administration provides training in specific areas for the members of EU staff; and
- a host of specific agencies and bodies handle a range of technical, scientific and management tasks.

The powers and responsibilities of all of these institutions are laid down in the treaties, which are the foundations of the EU formal decisions. The treaties also lay down the rules and procedures that the EU institutions must follow. They are agreed by the Presidents and/or Prime Ministers of all the EU countries and ratified by the national Parliaments of the Member States.

1.4 THE EU LEGISLATIVE PROCEDURES

In the Ordinary Legislative Procedure (see Figure 1.1) of the European Union, the Commission sends each primary legislative proposal to the

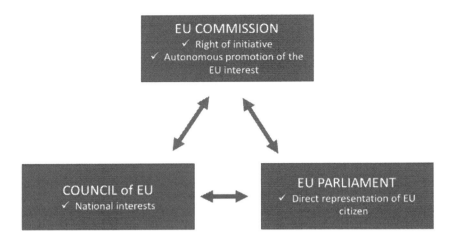

Figure 1.1   The Ordinary Legislative Procedure of the EU.

Parliament and the Council. They consider it and discuss it on two successive reading procedures, if one procedure is not sufficient to adopt an identical version of the legislative proposal. After two readings, if they do not agree, the proposal is brought before a Conciliation Committee made up of an equal number of representatives of the Council and the Parliament. Representatives of the Commission also attend the meetings of the Conciliation Committee and contribute to the discussions. When the Committee has reached an agreement, the text agreed upon is sent to the Parliament and the Council for a third reading. The final agreement of the two institutions on an identical text is essential if it has to be adopted as a law. Even if a joint text is agreed by the Conciliation Committee, it can still be rejected if the third reading is negative.

Figure 1.1 highlights, in particular, the role of the three main EU institutions in the ordinary regulatory process, whereas Figure 1.2 highlights the importance of the contribution provided by European Food Safety Agency for the adoption of EU food law and decision making through the adoption of scientific opinions.

It is also important to consider the existence in the EU of the delegated legislative procedure that is only applicable to specific Regulations whose adoption is delegated by primary legislation to a joint working group of the European Commission and representatives of Member States. This is

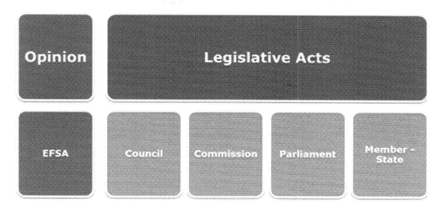

Figure 1.2   Decision making in EU food law.

the case of legislative acts for updating technical Regulations to scientific and technical developments such as the Regulations on additives, contaminants, enzymes and similar ones.

The Commission's power to adopt delegated acts is subject to strict limits:

- the delegated act cannot change the essential elements of the primary law;
- the primary legislative act must define the objectives, content, scope and duration of the delegation of power; and
- Parliament and Council may revoke the delegation or express objections to delegated acts.

The Commission prepares and adopts delegated acts after consulting expert groups, composed of representatives from each EU Member State, which meet on a regular or occasional basis.

As part of the Commission's better Regulation agenda, citizens and other stakeholders can provide feedback on the draft text of a delegated act during a four-week period. There are some exceptions, for example, in case of emergency or when citizens and stakeholders have already contributed comments.

Once the Commission and the representatives of the Member States have adopted a delegated act, Parliament and Council generally have two months to formulate any objections. If they do not object, the delegated act enters into force.

Adopted acts contain an "explanatory memorandum" summarizing the feedback received and how it was dealt with.

A new Interinstitutional Register of Delegated Acts was launched in December 2017 and is available in all the EU languages. It provides a complete view of the lifecycle of delegated acts and allows users to subscribe in order to receive notifications about the files of their interest.

2

Major Foodborne Illnesses and Causes

2.1 INTRODUCTION

Unsafe food may pose global health threats, potentially endangering any consumer. Infants, young children, pregnant women, the elderly and those with an underlying illness are particularly vulnerable. For many years, WHO has been monitoring worldwide foodborne illnesses which are usually infectious or toxic in nature and are caused by bacteria, viruses, parasites or chemical substances entering the body through contaminated food or water (WHO, World Health Statistics, 2015 and 2019). The Second International Conference on Nutrition (ICN2), held in Rome in November 2014, reiterated the importance of food safety in achieving better human nutrition through healthy nutritious diets. Improving food safety is thus a key target in achieving Sustainable Development Goals. The worldwide burden of foodborne diseases to public health and welfare and to the economy has often been underestimated due to underreporting and difficulty in establishing causal relationships between food contamination and the resulting illness or death.

2.2 FOOD SAFETY: A PUBLIC HEALTH PRIORITY

The 2015 WHO report on World Health Statistics provides the first ever estimates of the global burden of foodborne diseases caused by 31 foodborne agents (bacteria, viruses, parasites, toxins and chemicals) at worldwide

9

and regional levels. Foodborne pathogens can cause severe diarrhoea or debilitating infections, including meningitis. Chemical contamination can lead to acute poisoning or long-term diseases, such as cancer. Foodborne diseases may lead to long-lasting disability and death. Examples of unsafe food include uncooked foods of animal origin, fruits and vegetables contaminated with faeces, and raw shellfish containing marine biotoxins.

Food can become contaminated at any point of production and distribution, and the primary responsibility lies with food producers. Yet a significant proportion of foodborne disease incidents are caused by foods improperly prepared at home, in food service establishments or in markets. Therefore, all food handlers and consumers should fully understand the roles they must play, such as adopting basic hygienic practices when buying, selling and preparing food to protect their health and that of the wider community. Urbanization and changes in consumer habits, including travel, have increased the number of people buying and eating food prepared in public places. Globalization has triggered growing consumer demand for a wider variety of foods, resulting in an increasingly complex and longer global food chain. As the world's population increases, the intensification and industrialization of agriculture and animal production to meet an increasing demand for food creates both opportunities and challenges for food safety. Climate change is also predicted to impact food safety. For instance, temperature changes modify food safety risks associated with food production, storage and distribution. These challenges put greater responsibility on food producers and handlers to ensure food safety. Local incidents can quickly evolve into international emergencies due to the speed and range of product distribution. Serious foodborne disease outbreaks have occurred in every continent in the past and have been often amplified by globalized trade.

Examples include the contamination of infant formula with melamine in 2008 and the 2011 enterohaemorrhagic *Escherichia coli* outbreak linked to contaminated fenugreek sprouts, leading to some deaths and significant economic losses.

2.3 MAJOR FOODBORNE CAUSES OF ILLNESSES

2.3.1 Chemicals

Of most concern for health are naturally-occurring toxins and environmental pollutants.

- **Naturally occurring toxins** include mycotoxins, marine biotoxins, cyanogenic glycosides and toxins occurring in poisonous mushrooms. Staple foods like corn or cereals may contain high levels of mycotoxins, such as aflatoxin and ochratoxin, produced by mould on grain. Long-term exposure can affect the immune system and normal development or cause cancer.
- **Persistent organic pollutants (POPs)** are compounds that accumulate in the environment and the human body. Known examples are dioxins and polychlorinated biphenyls (PCBs), which are unwanted by-products of industrial processes and waste incineration. They are found worldwide in the environment and accumulate in the animal food chain. Dioxins are highly toxic and can cause reproductive and developmental problems, damage the immune system, interfere with hormones and cause cancer.
- **Heavy metals** such as lead, cadmium and mercury cause neurological and kidney damage. Contamination by heavy metal in food may occur mainly through pollution of air, water and soil (WHO, 2015).

2.3.2 Bacteria

- *Salmonella*, *Campylobacter* and **enterohaemorrhagic *E. coli*** are among the most common foodborne pathogens that may affect large numbers of people by causing fever, headache, nausea, vomiting, abdominal pain and diarrhoea. Examples of foods that have been involved in outbreaks of salmonellosis are eggs, poultry and other products of animal origin. Foodborne cases with *Campylobacter* have been mainly caused by raw milk, raw or undercooked poultry and drinking water. Enterohaemorrhagic *E. coli* has been associated with unpasteurized milk, undercooked meat and fresh fruits and vegetables.
- *Listeria* infection leads to the risk of unplanned abortions in pregnant women or the death of newborn babies. Although disease occurrence associated with Listeria exposure is relatively low, it may be quite severe and sometimes may have fatal health consequences, particularly among infants, children and the elderly. Listeria may be present in unpasteurized dairy products and various ready-to-eat foods and can grow at refrigeration temperatures.
- *Vibrio cholerae* may infect people through contaminated water or food and cause abdominal pain, vomiting and profuse watery

11

diarrhoea, which may lead to severe dehydration and possible death. Rice, vegetables, millet gruel and various types of seafood have been implicated in cholera outbreaks.

Antimicrobials, such as antibiotics, are essential to treat infections caused by bacteria. However, their overuse and misuse in veterinary and human medicine has been linked to the emergence and spread of resistant bacteria, rendering the treatment of some infectious diseases ineffective in animals and humans. Resistant bacteria enter the food chain through the animals (e.g. *Salmonella* through chickens). Antimicrobial resistance is one of the main threats to modern medicine (WHO, 2015).

2.3.3 Viruses

Norovirus infections are characterized by nausea, explosive vomiting, watery diarrhoea and abdominal pain. Hepatitis A virus can cause long-lasting liver disease and spreads typically through raw or undercooked seafood or contaminated raw produce. Infected food handlers are often the source of food contamination (WHO, 2015).

2.3.4 Parasites

Some parasites, such as fish-borne trematodes, are transmitted only through food. Others, for example, tapeworms like *Echinococcus* spp. or *Taenia solium*, may infect people through food or direct contact with animals. Other parasites, such as *Ascaris, Cryptosporidium, Entamoeba histolytica* or *Giardia*, enter the food chain via water or soil and can contaminate fresh produce (WHO, 2015).

2.3.5 Prions and BSE Crisis (Discovery and Evolution, Geographical BSE Risk Assessment and Risk Organ Removal and Other Control Tools)

Prions, infectious agents consisting of specific proteins, are unique in that they are associated with specific forms of neurodegenerative diseases (spongiform encephalopathy). Bovine spongiform encephalopathy, commonly known as mad cow disease, is a neurodegenerative disease of cattle characterized by symptoms such as abnormal behaviour, trouble walking and weight loss. Later in the course of disease development, the cattle becomes unable to move. The time between the infection and the onset of symptoms is generally four to five years. The time from the onset of symptoms to death is generally weeks to months (Silano, 2003). BSE has evolved rapidly since 1987 in the UK into an issue of major public concern due to the very large number of affected cattle (several thousands per year). Available data apparently suggest that reduction of incidence of BSE started in 1993 and became dramatic in 1999. However, confirmed BSE figures do not necessarily present an accurate picture of the disease as many cattle affected by BSE may have been slaughtered before the disease due to the fact that 4.4 millions of cattle have been slaughtered in the UK under a series of governmental schemes during the BSE eradication programme. Spread of BSE to humans was shown to result in variant Creutzfeldt–Jakob disease (vCJD). In 1996 evidence was provided that this disease had crossed the species barrier and infected humans in the UK (about 10 cases in 1996 and 1997, 20 in 1998, 15 in 1999 and 20 in 2000 were reported by the vCJD Surveillance Unit). vCJD is a degenerative brain disease in humans, which is thought to be caused by an abnormal prion protein in the brain. Its most likely origin is exposure to the BSE prions via eating tissues from infected BSE cattle. The number of deaths from definite or probable vCJD in the UK is estimated to be 178 (The national cjd research & surveillance unit)

BSE has become the most important international animal disease in the last 50 years, with a significant impact on the national and international management of animal diseases, in general, on economics, trade, and public health issues related to food.

The Geographical BSE Risk (GBR) methodology was developed by the EU Scientific Steering Committee in response to the demand from the EC to find a way to identify the occurrence of BSE and to deal with the risk of importing the BSE agent from countries that had not confirmed its presence through the identification of affected animals. The conventional way of dealing with this animal disease and its control based on diagnosed cases was, in fact, at that time not regarded as adequate for BSE, given the expected low prevalence and the many difficulties in testing the disease to detect the affected animals.

Until 1999, self-reporting of BSE cases was the basis for trade decisions. Widespread import bans followed the initial reports of BSE cases in new countries. Such dramatic import restrictions strongly demotivated some countries from developing a good BSE surveillance and reporting cases system. The World Organization for Animal Health (OIE) established early in the evolution of the Terrestrial Animal Health Code for BSE that the BSE status of a country could only be determined on the basis of a risk assessment. Therefore, a method to determine the BSE risk of countries was urgently needed.

In response to this challenge, in 1997 the European Commission Scientific Steering Committee (SSC) combined the available knowledge about the disease and established a model of the BSE/cattle system that explained and was consistent with the observed dynamics of the disease in the UK, where the most data were available. The Geographycal BSE Risk Assessment Methodology valuated qualitatively, but efficiently, the likelihood that an animal within a country or region was infected with BSE based on the probability that a BSE agent would enter the country (external challenge) and the probability that the agent would amplify and spread within a country (internal stability) (Salman et al., 2012).

Many countries found their first BSE cases and introduced important control measures only after the publication of their GBR status in the year 2000. This publication clearly showed that most of the countries in Europe and worldwide undergoing the evaluation were actually affected by BSE, although very often competent national authorities were convinced of the contrary. Obviously, the GBR credibility was significantly enhanced when the "soft", qualitative GBR predictions were confirmed by "hard", quantitative reliable surveillance data after the year 2000.

The active surveillance programmes finally gave a much better picture of the real incidence of this disease and showed how much better active surveillance can elucidate critical diseases having a very low prevalence if key risk factors are included.

BSE has been most often spread through the practice of feeding cattle various meats (rendered materials) from slaughtered animals, such as already infected cattle. During this process, an abnormal protein that is linked to BSE can spread from a slaughtered diseased animal to a healthy one. This abnormal protein, called "prion", can withstand high temperatures and does not get destroyed during the rendering procedure. Since the incubation period for BSE is so long, it is possible for an infected animal to enter the food chain before the symptoms appear. Proteins are complex molecules folded up into particular shapes. A prion is folded differently from a normal protein, and it can cause normal proteins to change and fold abnormally. The cells most often infected are the brain cells. The resulting solidification of the proteins causes the infected brain tissues to look like a sponge with several tiny holes, hence the name "spongiform encephalopathy" for this disease.

The best way to prevent the disease from spreading among cattle is to avoid feeding cattle rendered material from slaughtered animals and to isolate and destroy all infected animals. Currently, most countries have developed policies for monitoring BSE in their cattle herds and procedures for dealing promptly and thoroughly with BSE cases when they arise.

Overall, the BSE crisis, experienced during the past 20 years of the last century, clearly highlighted the substantial limitations of the food safety system at the time in place in the European Union and motivated drastic initiatives of the European Commission, known as "the white paper on food safety" and the "farm to table legislation" whose implementation has made possible to substantially improve the food safety level in the European Union (see Chapter 3).

3

The EU Mandate to Promote Food Safety and "The White Paper on Food Safety" and the "Farm to Table" Legislation

3.1 INTRODUCTION

Apart from the provision providing more general competence in the area of Agriculture (Article 43 TFEU) and the internal market (Article 114), respectively, the European Union Treaty offers two stepping stones for the European Union competence to harmonize in the field of consumer health protection. The Article 168(4) TFEU allows for harmonization in order to meet a number of "common safety concerns" for which food safety is relevant, whereas Article 169(2), in conjunction with Article 114 TFEU, states that the EU shall contribute to protecting the health and safety of food consumers, thus clearly identifying the EU competence in the sector of "food health" (Edinger, 2014). Relevant in the present context is an article (Arpaia, 2017) that outlines the path that led to the birth at the international level of the right to food safety by retracing the development and framing especially the turning point with the recall of the international food principles in World Trade Organization agreements.

The European Union works closely with WHO, FAO, the World Organization for Animal Health (OIE) and other international organizations to ensure food safety along the entire food chain from production to consumption.

3.2 THE EU MANDATE TO PROMOTE FOOD SAFETY

The implementation of the integrated Food Safety policy in the EU involves various actions, namely:

- to assure effective control systems and evaluate compliance with EU standards in the food safety and quality, animal health, animal welfare, animal nutrition and plant health sectors within the EU and in third countries in relation to their exports to the EU;
- to manage international relations with third countries and international organizations concerning food safety, animal health, animal welfare, animal nutrition and plant health; and
- to manage relations with the EFSA and ensure science-based risk management.

The integrated approach to food safety recognized that:

(i) European citizens have the right to know how the food on the market is produced, processed, packaged, labelled and sold;

(ii) the central goal of the European Commission's Food Safety policy is to ensure a high level of protection of human health regarding the food industry, which is the largest manufacturing and employment sector in Europe; and

(iii) the European Commission's guiding principle – primarily set out in its White Paper on Food Safety – is to apply an integrated approach from farm to fork covering all sectors of the food chain.

3.3 THE WHITE PAPER ON FOOD SAFETY AND THE "FARM TO TABLE" LEGISLATIVE ACTION

On 12 January 2000, the EC adopted the "White Paper on Food Safety" (European Commission, 2000). The central goal of the EC was the improvement of the level of health protection for food consumers in the European Union. The White Paper sets out proposals for a programme of legislative reform to complete the EU's "farm to table" approach, as well as the

18

establishment of the EFSA. Achieving the highest standards of food safety in the EU was a key policy priority for the EC, and the White Paper bears testimony to this priority. The guiding principle throughout the White Paper was that food safety policy must be based on a comprehensive and integrated approach. The Commission also decided on the allocation of food safety and industrial policy responsibilities.

Major initiatives addressed in the White Paper are described in the following sections.

3.3.1 The European Food Safety Authority

Among the weaknesses identified in the White Paper there were (i) the lack of scientific support for the system of scientific advice; (ii) the inadequacies in monitoring and surveillance on food safety issues; (iii) the gaps in the rapid alert system and lack of coordination of scientific co-operation and analytical support. The White Paper envisaged the establishment of the European Food Safety Authority based on the principles of independence, scientific excellence and transparency in its operations. Therefore, the Authority must be guided by the best science, be independent of industrial and political interests, be open to rigorous public scrutiny, be scientifically authoritative and work closely with national scientific bodies. The tasks of the new Authority were identified on risk assessment and risk communication. Risk management, including legislation and control, was expected to remain the responsibility of the European institutions, which are accountable to the European public. The Authority is expected to take a proactive role in developing and operating food safety monitoring and surveillance programmes. It will need to establish a network of contacts with similar agencies, laboratories and consumer groups across the EU and in third countries. Finally, the Authority will need to make special provision for informing all interested parties of its findings, not only in respect of the scientific opinions but also in relation to the results of its monitoring and surveillance programmes. The Authority is expected to become the first port of call when scientific information on food safety and nutritional issues is sought or problems have been identified. A highly visible Authority with a strong proactive presence on food safety matters was considered to be a key element in restoring and maintaining confidence among European consumers. According to the White Paper, the Authority should also collaborate with the rapid alert system to be significantly strengthened as part of this process and to include also rapid alert for animal feed problems.

3.3.2 Food Safety Legislation

The White Paper proposed an action plan with a wide range of measures to improve and bring coherence to the Community's legislation, covering all aspects of food products from "farm to table" with over 80 separate actions. The new legal framework was expected to cover animal feed, animal health and welfare, hygiene, contaminants and residues, novel foods, additives, flavourings, packaging and irradiation and to include a proposal on General Food Law to embody the principles of food safety such as:

- responsibility of feed manufacturers, farmers and food operators;
- traceability of feed, food and their ingredient;
- proper risk analysis through: (a) risk assessment (scientific advice and information analysis), (b) risk management (regulation and control), and (c) risk communication; and
- the application of the precautionary principle, as appropriate.

3.3.3 Control of Legislation Implementation

A comprehensive piece of legislation was proposed in order to recast the different control requirements and to take into account the general principle that all parts of the food production chain must be subject to official controls. A clear need for a Community framework of national control systems to improve the quality of controls at the Community level and consequently to raise food safety standards across the European Union was identified. The operation of such control systems would remain mainly a national responsibility. This Community framework would have three core elements:

- operational criteria set up at Community level;
- Community control guidelines; and
- enhanced administrative co-operation in the development and operation of control.

Development of this overall Community framework for national control systems was clearly indicated to be a joint task for the Commission and the Member States.

3.3.4 Consumer Information

The Commission, together with the new EFSA, was expected to promote a dialogue with consumers to encourage their involvement in the new food safety policy. At the same time, consumers need to be kept better informed of emerging food safety concerns and risks to certain groups from particular foods. Proposals on the labelling of foods, building on existing rules, are to be brought forward.

3.3.5 International Dimension and Other Aspects

As the European Community is the world's largest importer/exporter of food products, the actions proposed in the White Paper needed to be effectively presented and explained to trading partners. An active role for the Community in international bodies is an important element in explaining European developments in food safety.

On the day when the "White Paper" was adopted, it was also decided that the Commissioner responsible for Health and Consumer Protection should have the overall responsibility for all elements of the food chain. Accordingly, the European Commission decided on that day to unify all the competencies for food safety to the Directorate General of DG Health.

4

Regulation EC 178/2002
An Integrated Approach to Support Food Safety

4.1 INTRODUCTION

Regulation (EC) 178/2002 (also referred to as the General Food Law), as amended by Regulation (EC) 1624/2003 of the European Parliament and the Council and by Regulation (EC) 573/2006, has provided for about 18 years the basis for the assurance of a high level of protection of human health and consumers' interest in relation to food, taking into account, in particular, the diversity in the supply of food including traditional products, whilst ensuring the effective functioning of the internal market. It has established common principles and responsibilities, the means to provide a strong science base, efficient organizational arrangements and procedures to underpin decision-making in matters of food and feed safety. Sanctions applicable to the violations of Regulation (EC) 178/2002 were adopted in Italy with the Legislative Decree 130/2006 on 6 April 2006.

This food law aims to protect the interests of consumers and to provide a basis for consumers to make informed choices in relation to the foods they consume by preventing (i) fraudulent or deceptive practices, (ii) food adulteration and (iii) any other practices which may mislead the consumer. This Regulation applies to all stages of production, processing and distribution of food and feed but not to primary production for private domestic use or to the domestic preparation, handling or storage of food for private domestic consumption. According to Article 2 of this Regulation, "food" (or "foodstuff") means any substance or product, whether processed, partially processed or unprocessed, intended to be, or reasonably expected to be, ingested by human beings. "Food definition" includes drink, chewing gum and any substance, including water, intentionally incorporated into the food during its manufacture, preparation or treatment.

The last amendments of Regulation (EC) 178/2002 have been adopted on 6 September 2019 with the new Regulation 2019/1381 on the transparency and sustainability of the EU risk assessment in the food chain (see Chapter 12 of the present book) that entered into force 20 days later. However, these amendments only apply 18 months after entering into force with the exception of measures in Article 1(2) that apply only from 1 July 2022 and in Article 1(3) that apply from the date of the appointment of the members of the new EFSA Scientific Panels.

4.2 RISK AND RISK ANALYSIS IN THE FOOD CHAIN

In order to achieve the general objective of a high level of protection of human health and life, food safety should be based on risk and risk analysis. Risk is a function of the probability and the severity of an adverse health effect, consequential to a hazard occurring in the food chain.

The hazards (or risk factors) which may be associated with the food chain include:

- hazardous contaminants of biological, chemical and/or physical nature;
- naturally occurring substances of concern, especially if present as secondary metabolites in botanical and/or other products; and
- unbalanced diets and dietary habits promoting excessive or inadequate nutrient intakes.

Risk analysis, which is fundamental to efficiently control risk factors associated with the food chain, consists of:

- risk assessment carried out by qualified scientists through (i) hazard identification and characterization, (ii) exposure assessment, and (iii) risk characterization, all based on scientific evidence and excellence, independence, objectivity and transparency (see also Section 11.7);
- risk management, based on the results of risk assessment, to prevent or minimize the risk (in case of no compliance of their decisions with expert scientific advice, risk managers have to provide explanations); and

- risk communication to food consumers, which is not only a duty of risk managers but also of risk assessors.

Risk assessment shall be based on the available scientific evidence and undertaken in an independent, objective and transparent manner. Risk assessment and communication are the main tasks of EFSA. This implies providing scientific advice and scientific and technical support for the EU's legislation and policies in all fields which have a direct or indirect impact on food and feed safety, and providing independent information on all matters within these fields and communicating on risks.

For additional measures concerning EFSA's risk assessment, see Sections 11.7 and 11.8, and for their on-going evolution, see Chapter 12.

In their activities, risk managers shall take into account the results of risk assessment and, in particular, the scientific opinions provided by EFSA.

4.3 PRECAUTIONARY PRINCIPLE

In specific circumstances where, following an assessment of available information, the possibility of harmful effects on health is identified but scientific uncertainty persists, provisional risk management measures necessary to ensure the high level of health protection may be adopted, pending further scientific information for a more comprehensive risk assessment.

Control measures shall be proportionate, and no more restrictive of trade than is required to achieve the high level of health protection shall be chosen in the EU, having regard to technical and economic feasibility and other factors regarded as legitimate in the matter under consideration. The measures shall be reviewed within a reasonable period of time, depending on the nature of the risk to life or health identified and the type of information needed to clarify the scientific uncertainty and conduct a more comprehensive risk assessment.

4.4 TRANSPARENCY PRINCIPLES

There shall be an open and transparent public consultation, directly or through representative bodies, during the preparation, evaluation and revision of food law, except where the urgency of the matter does not allow it.

Where there are reasonable grounds to suspect that a food or feed may present a risk for human or animal health, then, depending on the risk nature, seriousness and extent, public authorities shall take appropriate steps to inform the general public of the nature of the risk. Moreover, there is a need for identifying to the fullest extent possible the food or feed, the risk that it may present and the measures which are taken or about to be taken to prevent, reduce or eliminate that risk.

Food and feed imported for being placed on the market within the European Union shall comply with the relevant requirements of food law or conditions recognized by the EU to be at least equivalent thereto or, where a specific agreement exists between the EU and the exporting country, with requirements contained therein, whereas food and feed exported or re-exported from the EU for being placed on the market of a third country shall comply with the relevant requirements of food law unless otherwise requested by the authorities of the importing country. In other circumstances, except in the case where foods are injurious to health or feeds are unsafe, food and feed can be exported or re-exported only if the competent authorities of the country of destination have formally agreed, after having been fully informed of the reasons for which and the circumstances in which the food or feed concerned could not be placed on the market in the EU.

4.5 FOOD SAFETY REQUIREMENTS

According to Article 14 of Regulation (EC) 178/2002, food shall not be placed on the market if it is "unsafe" because it is (i) injurious to health or (ii) unfit for human consumption.

In determining whether any food is unsafe, regard is due to (i) the normal conditions of the use of the food by the consumer and at each stage of production, processing and distribution; and (ii) the information provided to the consumer, including information on the label, or other information generally available to the consumer, concerning the avoidance of specific adverse health effects from a particular food or category of foods. In determining whether any food is injurious to health, regard shall be given (i) not only to the likely immediate, short-term and/or long-term effects of that food on the health of a person consuming it but also on subsequent generations; (ii) to the probable cumulative toxic effects; and (iii) to the particular health sensitivities of a specific category of consumers for which the food is intended. Where any unsafe food is part of a batch, lot or consignment of food of the same class or description, it shall be presumed that all the food in that batch, lot or consignment is also unsafe, unless following a detailed assessment that there is no evidence for such a conclusion.

4.6 FEED SAFETY REQUIREMENTS

Feed shall not be placed on the market or fed to any food-producing animal if it is unsafe. Feed shall be deemed to be unsafe for the intended use if it is considered to: (i) cause an adverse effect on animal health; and/or (ii) make the food derived from food-producing animals unsafe for human consumption. Where a feed which has been identified as not satisfying the feed safety requirement is part of a batch, lot or consignment of feed of the same class or description, it shall be presumed that all of the feed in that batch, lot or consignment is so affected, unless following a detailed assessment that there is no evidence to extend such a conclusion to the rest of the batch, lot or consignment.

4.7 OTHER GENERAL APPROACHES ADOPTED TO ENSURE FOOD SAFETY

Other approaches include:

- proportionality (i.e. among different measures to achieve a specific safety objective, the less restrictive one should be adopted);
- non-discrimination and coherence (i.e. comparable cases should be managed with similar approaches);
- re-examination (i.e. risk assessment and related management need to be re-assessed whenever new relevant data become available on a specific persisting risk);
- transparency (every European citizen has the right to know how the food available on the market is produced, processed, packaged, labelled and sold and should be informed on the possible risk to health); and
- precautionary principle (when the possibility of harmful effects on health is identified but scientific uncertainty persists, provisional risk management measures necessary to ensure the high level of health protection chosen in the European Union should be adopted).

4.8 RESPONSIBILITIES OF FOOD BUSINESS OPERATORS

Food business operators (from farm to fork) have important responsibilities for risk prevention and minimization in the food chain, both under normal and under urgent, emergency and crisis conditions, to ensure a high level of protection of human health and consumers' interests. In particular, they have:

- the responsibility to prevent that, through their activities in the food/feed chain, unsafe products enter the market;
- a key role, not only in complying with relevant safety Regulations but also in carrying out a risk assessment in relation to their activities;
- the responsibility of ensuring systems which allow traceability of food and feed products and to undertake, when needed, withdrawal and recall; and
- the duty of collaborating with risk managers in case of urgent, emergency or crisis response and of notifications through the Rapid Alert System in Food and Feed (RASFF).

4.9 TRACEABILITY APPROACH

According to Article 18 of Regulation (EC) 178/2002, the traceability of food, feed, food-producing animals and any other substance intended, or expected, to be incorporated in a food or feed shall be established at all stages of production, processing and distribution. Food and feed business operators shall be able to identify any person from whom they have been supplied (or that they have supplied) with a food, a feed, a food-producing animal, or any substance intended, or expected, to be incorporated into a food or feed. To this end, such operators shall have in place systems and procedures which allow for this information to be made available to the competent authorities on demand. Food and feed business operators shall have in place systems and procedures to identify the other businesses to which their products have been supplied. This information shall be made available to the competent authorities on demand.

Food or feed which is placed on the market or is likely to be placed on the market in the European Union shall be adequately labelled or identified to facilitate its traceability through relevant documentation or information in accordance with the relevant requirements of more specific provisions. Therefore, it is important that food business operators save carefully all the information useful for traceability and, especially, the:

- nature of the food product received or provided;
- amount expressed in weight or volume units;
- presentation status;

- name of the provider/client of the food product;
- batch number of the product or other identification system;
- reception/invoice dates; and
- identification of the carrier (if different from the provider).

The Commission Implementing Regulation (EU) 931/2011 lays down provisions implementing the traceability requirements set by Regulation (EC) 178/2002 to food business operators in respect of the food of animal origin. This Regulation applies to food defined as unprocessed and processed products in Article 2(1) of Regulation (EC) 852/2004 and does not apply to food containing both products of plant origin and processed products of animal origin.

According to Article 3 of this Regulation (traceability requirements), food business operators shall ensure that the following information concerning consignments of food of animal origin is made available to the food business operator to whom the food is supplied and, upon request, to the competent authority: (a) an accurate description of the food; (b) the volume or quantity of the food; (c) the name and address of the food business operator from which the food has been dispatched; (d) the name and address of the consignor (owner) if different from the food business operator from which the food has been dispatched; (e) the name and address of the food business operator to whom the food is dispatched; (f) the name and address of the consignee (owner), if different from the food business operator to whom the food is dispatched; (g) a reference identifying the lot, batch or consignment, as appropriate; and (h) the date of dispatch. The information referred to shall be made available in addition to any information required under relevant provisions of Union legislation concerning the traceability of the food of animal origin. The information referred to shall be updated on a daily basis and kept at least available until it can be reasonably assumed that the food has been consumed. When requested by the competent authority, the food business operator shall provide the information without undue delay. The appropriate form in which the information must be made available is up to the choice of the supplier of the food, as long as the information requested is clearly and unequivocally available to and retrievable by the business operator to whom the food is supplied.

4.10 SYSTEM OF TRACEABILITY (TRACES)

According to the European Commission website available online, the European Commission requires consignments of animals, semen and embryo, food, feed and plants to be accompanied by health certificates or trade documents. When such consignments are exported to the EU or traded within the EU single market, TRACES (Trade Control and Export System) manages the official controls and route planning quickly and efficiently online. National competent authorities post these documents online through TRACES, and either the EU border control authorities or the control authorities at the destination check the consignments and their accompanying certificates to allow them to enter and travel through the EU.

National competent authorities are pre-notified of the arrival of a consignment and, therefore, can plan their controls. In the case of the transport of animals, further checks and controls on animal welfare can be performed. Feedback on the decision and controls is available to those parties concerned with the consignment. With its user-friendly interface, available in 35 languages, shared by the private sector and the competent authorities, TRACES relieves much of the administrative burden usually involved in these processes. Benefits of using TRACES for traders, consumers, competent authorities, animals and plants are described in detail in the European Commission website.

Therefore, TRACES is an efficient tool to ensure:

- **traceability** (monitoring movements, both within the EU and from non-EU countries);
- **information exchange** (enabling trade partners and competent authorities to obtain information on the movements of their consignments and speeding up administrative procedures); and
- **risk management** (reacting rapidly to health threats by tracing the movements of consignments and facilitating the risk management of rejected consignments).

TRACES aims to strengthen cooperation with EU partners, facilitate trade, accelerate administrative procedures and improve the risk management of health threats while combating fraud and enhancing the safety of the food chain, animal health and plant health.

4.11 MORE POWERS FOR THE EUROPEAN COMMISSION TO PREVENT RISKS UNDER EMERGENCY CONDITIONS

Where it is evident that food or feed originating in the Union or imported from a third country is likely to constitute a serious risk to human health, animal health or the environment and that such risk cannot be contained satisfactorily by means of measures taken by the Member State(s) concerned, the Commission, acting in accordance with the procedure provided for in Article 58(2) of Regulation (EC) 178/2002, in the case of food or feed of Community origin, shall immediately adopt one or more of the following measures, depending on the gravity of the situation:

(i)   suspension of the placing on the market or use of the food/feed in question;
(ii) laying down special conditions for the food or feed in question; and
(iii) any other appropriate interim measure.

In the case of food or feed imported from a third country, the measures to be adopted, depending on the gravity of the situation, may include:

(i)   suspension of imports of the food or feed in question from all or part of the third country concerned and, where applicable, from the third country of transit;
(ii) laying down special conditions for the food or feed in question from all or part of the third country concerned; and/or
(iii) any other appropriate interim measure.

The Commission shall draw up, in close cooperation with EFSA and the Member States, a general plan for crisis management in the field of the safety of food and feed to specify the types of the situation involving direct or indirect risks to human health deriving from food and feed which are not likely to be prevented, eliminated or reduced to an acceptable level by provisions in place or cannot adequately be managed solely by way of the application of Articles 53 and 54 of Regulation (EC) 178/2002. The general plan shall also specify the practical procedures necessary to manage a crisis, including the principles of transparency to be applied and a communication strategy.

Where the Commission identifies a situation involving a serious direct or indirect risk to human health deriving from food and feed and the risk cannot be prevented, eliminated or reduced by existing provisions or cannot adequately be managed solely by way of the application of Articles 53 and 54 of Regulation (EC) 178/2002, it shall immediately notify the Member States and the EFSA and shall set up "a crisis unit" immediately, in which the EFSA shall participate to collect and evaluate all relevant information and identify the options available to prevent, eliminate or reduce to an acceptable level the risk to human health as effectively and rapidly as possible.

4.12 THE FITNESS CHECK OF THE GENERAL FOOD LAW (GFL)

On 15 January 2018, the Commission completed the fitness check on the General Food Law Regulation (Regulation (EC) 178/2002), which was launched in 2014 (European Commission, 2018). It is a comprehensive policy evaluation assessing whether the legislative framework introduced by the General Food Law Regulation for the entire food and feed sector is "fit for purpose" and whether it captures and reflects policy trends of today. The fitness check focuses on the General Food Law Regulation, which is the foundation of all legal measures at EU and national level in the area of food law. However, given its nature as framework legislation, this evaluation has also assessed the implementation of the common definitions, objectives, general principles and requirements set out in the General Food Law Regulation in other EU sectoral food and feed legislation. The fitness check has covered the period 2002–2013 in the EU 28 Member States. The mandate for the fitness check on the General Food Law Regulation, published in 2014, defines the overall scope and aim of the exercise and sets out the key questions to be addressed in relation to the fitness check criteria:

- effectiveness (Have the objectives been met?);
- efficiency (What are the costs and benefits involved?);
- coherence (Does the policy complement other actions, or are there contradictions?);
- relevance (Is the EU action still relevant?); and
- EU-added value (Can or could similar changes be achieved at the national/regional level, or did EU action provide clear added value?).

The collection of evidence, data and information constituted a critical part of this fitness check. To support the evidence-base of the fitness check, the European Commission procured two external studies launched in September 2014 and completed in December 2015 on the:

- general part of General Food Law Regulation (Articles 1–21); and
- RASFF and the management of emergencies/crises (Articles 50–57).

Broad stakeholder consultations (MS, stakeholders, etc.) have been performed during the whole fitness check exercise to collect the views of relevant actors in the food chain and gather information and evidence.

The main findings of the fitness check are:

- the General Food Law Regulation is still relevant today with respect to the current trends: growth and competitiveness and increased globalization. Nevertheless, it is less adequate to address new challenges like food sustainability in general, and more specifically, food waste;
- overall, the General Food Law Regulation has achieved its core objectives, namely high protection of human health and consumers' interests and the smooth functioning of the internal market;
- no systemic failures have been identified;
- current food safety levels are more favourable than before the adoption of the General Food Law Regulation;
- the systematic implementation of the risk analysis principles in EU food law has overall raised the level of protection of public health;
- the creation of EFSA has improved the scientific basis of EU measures. Major improvements in increasing EFSA's scientific capacity of expertise, the quality of its scientific outputs, its collection of scientific data and the development and harmonization of risk assessment methodologies have taken place;
- better traceability of food and feed in the entire agri-food chain;
- better transparency of the EU decision-making cycle;
- EU emergency measures and existing crisis management arrangements have overall achieved consumer health protection and the efficient management and containment of food safety incidents; and
- the General Food Law Regulation has contributed to the effective functioning of the internal market by creating a level playing field for all feed and food business operators in the EU market and reducing disruptions of trade where problems have occurred. The value of the EU internal trade in the food and drink sector has increased by 72% over the past decade. It has also contributed to the EU product safety recognition worldwide and to an improved quality perception for EU products in non-EU markets. The EU food and drink industry has achieved a more globally competitive position since 2003 vis-à-vis the main trading partners.

The following shortcomings have also been identified:

- national differences in the implementation and enforcement of the EU legislative framework, mainly occurring on a case-by-case basis, have been observed;

- despite considerable overall progress, transparency of risk analysis remains an important issue in terms of perception:
  - as regards risk assessment in the context of authorization dossiers, EFSA is bound by strict confidentiality rules and by the legal requirement to primarily base its assessment on industry studies, laid down in the GFL Regulation and in the multiple authorization procedures in specific EU food legislation. These elements lead civil society to perceive a certain lack of transparency and independence, having a negative impact on the acceptability of EFSA's scientific work by the general public;
  - risk communication has not always been effective with a negative impact on consumers' trust and on the acceptability of risk management decisions;
- a number of negative signals have been identified on the capacity of EFSA to maintain a high level of scientific expertise and to fully engage all MS in scientific cooperation; and
- lengthy authorization procedures in some sectors (e.g. feed additives, plant protection products, food improvement agents, novel foods, health claims) slow down the market entry process.

To overcome the above-mentioned shortcomings, the European Institutions have recently adopted a new Regulation, amending Regulation (EC) 178/2002, to increase the transparency and sustainability of the EU risk assessment. This is Regulation (EU) 2019/1381 of the European Parliament and the Council on the transparency and sustainability of the EU risk assessment in the food chain and amending Regulations 178/2002, 1829/2003, 1831/2003, 2065/2003, 1935/2004, 1331/2008, 1107/2009, and 2015/2283 and Directive 2001/18/EC (see Chapter 12 of the present book). This Regulation applies after 18 months since its entry into force, whereas specific sections (i.e. Articles 1(2) and 1(3)) apply later. It should be noted that the status of health in the European Commission was also monitored through the Eugloreh project (Silano, 2009).

5

Consumer's Information Regulations

5.1 INTRODUCTION

"Nutrition claim" is any claim stating, suggesting or implying that a food or one of its components may have particular beneficial nutritional properties. Health claims (general function health claims or claims referring to reduction of risk of disease or to children's development and health) are messages that state, suggest or imply a relationship between a food or one of its components and human health (Silano et al., 2013). Relevant in the context of mandatory and voluntary consumer's information is the "Study on the Impact of Food Information on Consumers' Decision Making" completed by the TNS European Behaviour Studies Consortium in December 2014.

5.2 MANDATORY CONSUMER'S INFORMATION

Regulation (EU) 1169/2011 deals mainly with the provision to consumers of mandatory information on food. This Regulation entered into force on 13 December 2014, but the obligation to provide nutrition information has been applied only since 13 December 2016.

5.2.1 Food Information *versus* Food Labelling

This Regulation considered a number of future developments to be adopted through as many as 17 mandatory and 18 optional future delegated acts in relation to the issues such as:

- the use of pictograms or symbols instead of words and numbers;
- adoption of readability rules for the label;
- indication of origin for specific foods; and
- specifications on the declarations of nutritional values.

Table 5.1 provides a global vision of the structure of this Regulation.

This Regulation applies to food business operators at all stages of the food chain, where their activities concern the provision of food information to consumers, to all foods intended for the final consumer, including foods delivered by mass caterers, and to foods intended for supply to mass caterers. This Regulation also applies to catering services provided by transport undertakings, when the departure takes place on the territories of the Member States to which the treaties apply.

The provision of food information as regulated by Regulation (EU) 1169/2011 pursues a high level of protection of consumers' health and interests by providing a basis for final consumers to make informed choices and for the safe use of food. Food information law also aims to achieve in the EU the free movement of legally produced and marketed food, taking

Table 5.1 Global Vision of Regulation (EU) 1169/2011

Chapter I	General measures Scope of applications/definitions
Chapter II	General principles Objectives and principles that regulate mandatory information of foods
Chapter III	General requirements Loyal practices of information and responsibilities
Chapter IV	Mandatory information: (i) Section I: content and presentation; (ii) Section II: detailed measures; and (iii) Section III: nutritional declaration
Chapter V	Voluntary information
Chapter VI	National measures
Chapter VII	Conclusive measures

into account, where appropriate, the need to protect the legitimate interests of producers and to promote the production of quality products.

Where mandatory food information is required, it shall concern information that falls, in particular, into one of the following categories:

(a) information on the identity and composition, properties or other characteristics of the food;

(b) information on the protection of consumers' health and the safe use of food, in particular, on:
   • compositional attributes that may be harmful to the health of certain groups of consumers;
   • durability, storage and safe use; and
   • the health impact, including the risks and consequences related to harmful and hazardous consumption of food;

(c) information on nutritional characteristics so as to enable consumers, including those with special dietary requirements, to make informed choices.

When considering the need for mandatory food information and enabling consumers to make informed choices, account shall be taken of a widespread need on the part of the majority of consumers for certain information to which they attach significant value or of any generally accepted benefits to the consumer.

5.2.2 More Loyal Information Practices

Food information shall not be misleading, particularly:

(a) as to the characteristics of the food and, in particular, as to its nature, identity, properties, composition, quantity, durability, country of origin or place of provenance, method of manufacture or production;

(b) by attributing to the food effects or properties which it does not possess;

(c) by suggesting that the food possesses special characteristics when in fact all similar foods possess such characteristics, in particular, by specifically pointing on the presence or absence of certain ingredients and/or nutrients; and

(d) by suggesting, by means of the appearance, the description or pictorial representations, the presence of a particular food or an ingredient, while in reality a component naturally present or an ingredient normally used in that food has been substituted with a different component or a different ingredient (Table 5.2).

Table 5.2 Some Key Aspects of Loyal Food Information

It should not be suggested that a food product has special characteristics if they are shared by all similar food products; this is particularly important with reference to the presence/absence of particular ingredients

It should not be suggested the presence of a specific food or ingredient when a natural component has been actually substituted

The food information, publicity and presentation (including form, appearance, packaging and way of the exhibition) must not be misleading but accurate, clear and simple to be understood

Mandatory food information should have precedence on voluntary food information

Food information shall be accurate, clear and easy to understand for the consumer. Subject to derogations provided for by the Union law applicable to natural mineral waters and foods for particular nutritional uses, food information shall not attribute to any food the property of preventing, treating or curing a human disease, nor refer to such properties.

5.2.3 Responsibilities Clarification

The food business operator responsible for the food information is the operator under whose name or business name the food is marketed or, if that operator is not established in the Union, the importer into the Union market.

The food business operator responsible for the food information shall ensure the presence and accuracy of the food information in accordance with the applicable food information laws and requirements of relevant national provisions.

Food business operators, within the businesses under their control, shall not modify the information accompanying a food if such modification would mislead the final consumer or otherwise reduce the level of consumer protection and the possibilities for the final consumer to make informed choices. Food business operators are responsible for any changes they make to food information accompanying a food.

Food business operators, within the businesses under their control, shall ensure that information relating to non-prepacked food intended for the final consumer or for supply to mass caterers shall be transmitted to the food business operator receiving the food in order to enable, when required, the provision of mandatory food information to the final consumer.

Food business operators that supply to other food business operators food not intended for the final consumer or to mass caterers shall ensure that those other food business operators are provided with sufficient information to enable them, where appropriate, to meet their obligations.

In the business-to-business sale, for pre-packaged food products ready for consumption, but also commercialized to catering companies for the subsequent preparation, separation or transformation, most of the information can be transferred through documents and only the following data must be present on the external packaging: name, duration, special conditions of storage or use and details concerning name and address.

Mandatory food information shall appear in a language easily understood by the consumers of the Member States where the food is marketed. Within their own territory, the Member States in which a food is marketed may stipulate that the particulars shall be given in one or more languages from among the official languages of the Union. The particulars can also be indicated in several languages.

5.2.4 Mandatory Food (Label) Information

The indications of the following particulars are mandatory:

   (i) the name of the food;
  (ii) the list of ingredients;
 (iii) any ingredient or processing aid listed in Annex II of Regulation (EU) 1169/2011 or derived from a substance or product listed in the previously mentioned Annex II causing allergies or intolerances used in the manufacture or preparation of a food and still present in the finished product, even if in an altered form;
  (iv) the quantity of certain ingredients or categories of ingredients;
   (v) the net quantity of the food;
  (vi) the date of minimum durability or the "use by" date;
 (vii) any special storage conditions and/or conditions of use;
(viii) the name or business name and address of the food business operator referred to in Article 8(1) of Regulation (EU) 1169/2011;
  (ix) the country of origin or place of provenance where provided for according to Article 26 of Regulation (EU) 1169/2011;

(x) instructions for use where it would be difficult to make appropriate use of the food in the absence of such instructions;

(xi) with respect to beverages containing more than 1.2% by the volume of alcohol, the actual alcoholic strength by volume; and

(xii) the nutrition declaration.

The previously referred particulars shall be indicated with words and numbers. They may additionally be expressed by means of pictograms or symbols. Where the Commission adopts delegated and implementing acts, the particulars may alternatively be expressed by means of pictograms or symbols instead of words or numbers.

Mandatory food information shall be available and shall be easily accessible for all foods. In the case of prepacked food, mandatory food information shall appear directly on the package on a label attached thereto.

In the case of non-prepacked food, the provisions of Article 44 of Regulation (EU) 1169/2011 shall apply. Without prejudice to the national measures adopted under Article 44(2) of Regulation (EU) 1169/2011, mandatory food information shall be marked in a conspicuous place in such a way as to be easily visible, clearly legible and, where appropriate, indelible. It shall not in any way be hidden by, obscured by, detracted from or interrupted by any other written or pictorial matter or any other intervening material.

Without prejudice to specific Union provisions applicable to particular foods, when appearing on the package or on the label attached thereto, the mandatory particulars listed in Article 9(1) of Regulation (EU) 1169/2011 shall be printed on the package or on the label in such a way as to ensure clear legibility, in characters using a font size where the x-height, as defined in Annex IV, is equal to or greater than 1.2 mm. In case of packaging or containers, the largest surface of which has an area of less than 80 cm², the x-height of the font size shall be equal to or greater than 0.9 mm (Figure 5.1).

The particulars listed in points (i), (v) and (ix) of the previously mentioned Article 9(1) shall appear in the same field of vision.

5.2.5 Requirements for Distance Sale

"Distance sale" covers any contract concluded between a provider and a client without their simultaneous physical presence, e.g. through the internet or a catalogue. In such case all the mandatory information must

Minimum height 1,2mm for manufactures with surface area equal or more than ≥80 cm²
Minimum height 0,9mm for manufactures with surface area more than < 80 cm²

Figure 5.1 Dimension of characters in labelling.

be available before the contract is stipulated and no additional costs have to be charged because of the distance sale.

Without prejudice to the information requirements laid down in Article 9 of Regulation (EU) 1169/2011, in the case of prepacked foods offered for sale by means of distance communication

(a) mandatory food information, except the particulars provided in point (f) of Article 9(1) of Regulation (EU) 1169/2011, shall be available before the purchase is concluded and shall appear on the material supporting the distance selling or must be provided through other appropriate means clearly identified by the food business operator. When other appropriate means are used,

   (i) the mandatory food information shall be provided without the food business operator charging consumers' supplementary costs; and

   (ii) all mandatory particulars shall be available at the moment of delivery.

In the case of non-prepacked foods offered for sale by means of distance communication, the particulars required under Article 44 of Regulation (EU) 1169/2011 shall be made available in accordance with paragraph 1 of this article. Point (a) of paragraph 1 of this article shall not apply to foods offered for sale by means of automatic vending machines or automated commercial premises.

Some additional forms of expression and presentation are also considered by Regulation (EU) 1169/2011 (see Table 5.3).

The information that is mandatory to find in the labels of food products is listed in Table 5.4.

Table 5.3 Additional Mandatory Forms of Expression and Presentation

High content of caffeine and origin of vegetable oils and fats

Frozen meat and fish (freezing date or first freezing)

Food imitation

Added water and proteins

Pre-formed foods from meat and fish pieces

Table 5.4 The Mandatory Label Information for Food Products

Name of the food (*)

Ingredients list

Allergens

QUID (if needed)

Net quantity of food (*)

Date of minimum durability or expiry date ("use by" date)
Use modalities (in case a proper use could otherwise be difficult)

The nation of origin or place of provenience (if foreseen)

Nutritional declaration

Alcoholic content by volume for beverages containing more than 1.2% by volume of alcohol (*)

Batch number (mandatory – Dir 2011/91 – is necessary in case the date of minimum durability or the "use by" date do not provide, in the order, a clear indication of the day and month of manufacture)

- -

(*) In the same vision camp, that is all surface that can be seen from a single viewpoint. The occurrence in the same vision camp does not apply to reusable glass bottles or when the largest surface is less than 10 cm.

5.2.6 Food Denomination

The food name needs to be accompanied by additional information if there are conditions or treatments whose omission may be misleading:

- specific cases have emerged for the need to add the term "thawed" if the finished food product has been frozen and de-frozen before the sale and if (i) the freezing was not a technologically needed step; and (ii) the freezing might have an impact of the food safety or quality; and

- in case one food component has been substituted partially or completely (e.g. a hamburger containing a portion of vegetable texturized proteins or a pizza produced with a cheese analogue), it is necessary to provide an indication close to the name of the product, with a character corresponding at least to 75% of the name and at minimum equal to 1.2 mm (Figure 5.1).

For meat products and preparations and for fish products:

- every protein added as such, including the hydrolyzed protein of a different animal origin, must be indicated in the food name (extended requirement);
- the products obtained with different pieces to look like a single piece must be labelled as "formed meat" or "formed fish";
- for products looking like cuts, joints, slices, portions or fillet of the whole product, the water, if present in an amount above 5%, must be mentioned in the name of the product (this prescription extends a requirement already existing for fish and reduces the level of 10%, already existing for treated meat); and
- If the gut used in food manufacture is not edible, it is necessary to provide an indication for that.

5.2.7 Ingredients List

Engineered nanomaterials have to be indicated with the word "nano". An exemption exists for water removed up to 5% from meat, meat preparations, untreated fishery products and bivalve molluscs. The generic names "vegetable oil" and "vegetable fat" have been eliminated.

It is necessary to list in parenthesis the different vegetable oil species (e.g. soya, palm, and sunflower respecting the amounts in weight). Moreover, it must be declared whether the oil or the fat is "completely hydrogenated" or "partially hydrogenated".

5.2.8 Allergens

The Regulation 1169/2011 provides in Annex II the list of 14 substances or products causing allergies or intolerances which are subject to labelling as allergens. Allergens have to be clearly identified in the list of ingredients:

- through a different character type or style or colour;
- for each allergenic derivative, even if forms of the same allergen (as e.g. "casein (milk) or serum proteins (milk));

- in case there is not an ingredient list, it should be specified "contains …" (previously foreseen only for alcoholic beverages, but now extended to all products without the list of ingredients).
- In case the food or beverage name already mentions the allergen, no other information is necessary.

In case the allergens are present in non-prepackaged food as is generally the case in restaurants, fast food shops, bread or delicatessen shops, hospitals and schools, national regulations should be adopted to define the tools (e.g. menu, catalogues, manifests and other audio-visual tools) to be mandatorily used to make easily accessible for the consumers the information on their presence in food.

Even in the absence of the adoption of national regulations, Regulation (EU) 1169/2011 rules are applicable to non-pre-packaged foods. Therefore, the information must be easily visible and indelible, meaning that information in Annex II of Regulation (EU) 1169/2011 must be made available in writing until the Member States do not otherwise regulate this area at the national level.

5.2.9 Net Quantity Declaration

This declaration must be made on the basis of volume units for liquids and mass units for solids. There are exclusions for:

- products that lose considerable volume or weight and are sold on the basis of number or that are weighed in the presence of the seller;
- products with amounts lower than 5 g or 5 ml, other than spices or herbs;
- products sold on the basis of the declared number unless it is possible to count them (e.g. eggs);
- products sold candied or otherwise recovered; the net quantity should not include the covering.

5.2.10 Duration Indications

Duration indications that are exempted from the declaration are:

- plants and fresh fruits unpeeled/uncut;
- beverages with alcohol by volume >10%;
- jams of flour and bread intended to be consumed within 24 hours from the preparation;
- salt and vinegar;
- solid sugar products, jams and chewing gums;

For the indication "to be consumed within …", the date must be reported on each pre-packaged portion. Moreover, beyond that date, food shall be considered unsafe and should not be marketed.

5.2.11 Batch Codex

A batch codex should be present on the label to:

- identify a group of animals;
- connect with a daily production; or
- be linked with a traceability system.

5.2.12 Mandatory Alerts Applicable to Specific Food Products

There are also some mandatory alerts only applicable to specific food products. This is the case for:

- beverages with a high caffeine content (more than 150 mg/l): "High caffeine content. Not recommended for children or for pregnant or lactating women (caffeine content expressed as mg/100 ml)";
- solid foods to what caffeine has been added with a physiological objective (no threshold): "Contain caffeine. Not recommended for children and pregnant women (followed by the caffeine content in mg/100 g)";
- for frozen meat, frozen meat preparations and frozen untreated fish products: "Date of freezing or date of the first freezing if frozen more than one time with the indication 'frozen on day, month and year or frozen in (place of freezing) on day, month and year'".

5.2.13 Readability Criteria

Regulation (EU) 1169/2011 has introduced a minimum dimension of the characters in the label (Figure 5.1):

Manufactures with a surface area below < 25 cm² are exempted by the nutritional declaration.

Manufactures with a surface area below <10 cm² need only to report:

- food name;
- allergens;
- quantity; and
- expiry date/date of minimum durability.

The Regulation requires that the following information should appear in the same vision field:

(i) food name;
(ii) quantity and, where applicable; and
(iii) alcoholic content.

Press agencies can certify the conformity of the "packaging label" to relevant Regulations (Figure 5.2).

5.2.14 Mandatory Nutrition Information

The nutritional declaration is mandatory since 13 December 2016. Absolute quantities in 100 g or 100 ml of pre-packaged food products must be indicated in whatever part of the package by using a character of the foreseen size. The format should be tabular, but if the available room is inadequate, it can also be linear. The order of nutrients for the labelling in the back of the pack is shown in Figure 5.3. Besides the nutritional facts, the BOP has to report the net weight, lot number, ingredient list, denomination of the producer, site of production and expiry date or minimum term of conservation.

The nutritional declaration may be expressed as previously described and also as percentages of the reference intakes per 100 g/ml by systematically mentioning the reference.

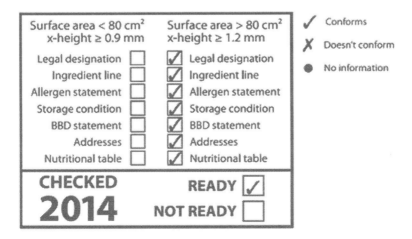

Figure 5.2 Press agencies can certify the conformity of the packaging.

NUTRITIONAL LABELLING	
NEW	**OLD**
Energy	Energy
Fats	Protein
Satured fats	Fats
Carbohydrates	Satured fats
Sugars	Carbohydrates
Protein	Sugars
Salt	Fiber
	Sodium

NUTRIENT (amount in 100 mg/ml and % of RDA per serving)	
MANDATORY	
Energy (Kcal/kJ)	
Fats (gr)	
Satured fats (gr)	
Carbohydrates (gr)	
Sugars (gr)	
Protein (gr)	
Salt (gr)	
OPTIONAL	
Fibre	
Mono-unsatured fats	
Poly-unsatured fats	
Polyols	
Starch	
MANDATORY FOR FUNCTIONAL FOODS (amount in 100 mg/ml and % of NRV per serving)	
Added vitamins and minerals (mg/ug)	

Figure 5.3   Back-of-the-pack nutritional labelling.

The average reference intakes of an adult are assumed to be, in terms of total energy: 8,400 kJ corresponding to 2,000 kcal; total fat: 70 g; saturated fats: 20 g; carbohydrates: 269 g; proteins: 50 g; and salt 6 g.

Facultative nutrients include:

- mono-unsaturated (g);
- poly-unsaturated (g);
- polyols (g) (more than two hydroxylic groups);

- starch (g);
- fibre (g); and
- anyone of the 13 vitamins and 14 minerals present in "significant amounts" defined as follows: for each package containing only one portion (e.g. some ready meals, beverages in 330 ml packages) at least 15% of the respective reference nutritional value in the single portion.

Otherwise, at least:

- 7.5% of the nutritional reference value in 100 ml for beverages;
- 15% of the nutritional reference value in 100 g/100 ml for products other than beverages.

The amounts of voluntarily added nutrients must be expressed per 100 g per 100 ml and may also be expressed for the portion or consumption unit. As far as the percentages of reference assumption dates: (i) for vitamins and minerals, it must be provided for 100 g/100 ml, and it may also be given for portion/consumption unit; and (ii) it cannot be given for the other added nutrients.

When the nutritional declaration has been provided at the back of the pack, it is also helpful to repeat the following information in the main vision field of the front of the pack (Figure 5.4):

- energy, or
- energy, fats, saturates, sugars and salt.

The main vision field is the most likely area to be noted at first glance from a consumer to make possible the identification of the character and nature of the product. Several such vision fields may exist.

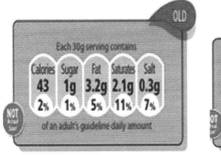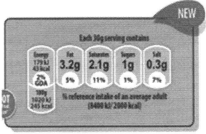

Figure 5.4   FoP labelling.

When the nutritional information is reported in the front of the pack, it is mandatory that the energy is expressed for 100 g/100 ml in both KJ and kcal. The nutrients (and their percentages) may be expressed for 100 g/100 ml and/or portion.

For the labelling of non-pre-packaged foods, the Member States may continue to apply their regulations.

In case of the absence of national regulations, the voluntary nutritional labelling can only include:

- the value of energy; or
- the value of energy and the amounts of fats, saturated fats, sugars and salt.

The value of energy is generally obtained through calculations, whereas for ingredients average values are derived through *ad hoc* analysis.

In case the values are negligible, the information can be substituted by the sentence "Contains negligible quantities of …". Negligible contents are:

- no more than 0.5 g/100 g or ml for fat, sugars, proteins and carbohydrates;
- no more than 0.1 g/100 g or ml for saturated fats; and
- no more than 0.005 g/0.0125 g for 100 g or ml for sodium/salt.

Other sectors where the Member States may adopt national regulations include:

- milk and milk products in glass bottles intended for re-use;
- ingredient lists in case of beverages containing more than 1.2% in the volume of alcohol; and
- voluntary indications of reference intakes for specific groups of population.

5.2.15 Additional Forms of Expression and Presentation

The European Commission has been charged with ensuring a "Community Register" of food nutrition and health claims. The "5 Icon System", used in the UK and Ireland, continues to be used by respecting the new order of presentation, for the labelling of the "Front of the Pack". Some years ago, the 5 Icon Systems has been proposed by the UK also in a three-colour version, where the red colour is used to identify nutrients considered to be present in excessive amounts as compared to the recommended ones .(Figure 5.5).

A 5-colours system (Nutri-Score) has been notified to EC by a draft Order the French Ministry of Social Affairs and Health on 24 April 2017 as the official (but voluntary) additional front-of-pack (FoP) nutrition label, which France is recommending to food business operators (FBOs) to promote healthier food choices. The Annex to the draft order describes the methods for calculating the unique overall score of each food based on the nutrient content to be based on the front of the pack. The relevant food category is shown on the "Nutri-Score" logo, which consists of five colours, each displaying a different letter from the best nutritional value to the worst (i.e. from dark green A, light green B, yellow C, light orange D to dark orange E) (Carreno, 2017) (Figure 5.5).

It is interesting to note that the possibility to make use of the voluntary information tools previously addressed derives directly from Article 35(2) of Regulation (EU) 1169/2011 that clearly states that the Member States recommend to FBOs the use of one or more additional forms of expression or presentation of the nutrition declaration that they consider as best fulfilling the following requirements (Carreno, 2017):

a) they are based on sound and scientifically valid consumer research and do not mislead the consumer;
b) their development is the result of consultation with a wide range of stakeholder groups;
c) they aim at facilitating consumer understanding of the contribution or importance of the food to the energy and nutrient content of a diet;
d) they are supported by scientifically valid evidence of the understanding of such forms of expression or presentation by the average consumer;
e) in the case of other forms of expression, they are based either on the harmonized reference intakes set out in Annex XIII of the mentioned regulation, or in their absence, on generally accepted scientific advice on intakes of energy or nutrients;
f) they are objective and not discriminatory; and
g) their application does not create obstacles to the free movement of goods.

Member States have to ensure appropriate monitoring of the additional forms of expression or presentation of the nutrition declaration that are present in the market of their territory.

By 13 December 2017, the European Commission was expected to submit a report to the European Parliament and to the Council on the use

Nutriscore

Colored 5 icon-systems

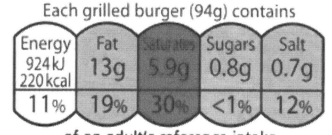

Figure 5.5 Two Supplementary voluntary FoP icons presented by France (Nutri-score) and UK to the EC (colored 5 icon system).

of additional forms of expression and presentation on their effect on the internal market and on the advisability of further harmonization of these forms of expression and presentation. In order to ensure the uniform application of Article 35 of Regulation (EU) 1169/2011, the Commission shall adopt implementing acts setting out detailed rules concerning the paragraphs 1, 3 and 4 of this article.

5.3 VOLUNTARY CONSUMER'S INFORMATION

Legislation on consumer's voluntary information has received a very high priority in the European Union (Paoletti et al., 2010; Silano, 2014a).

Regulation (EC) 1924/2006 on nutrition and health claims on food and food ingredients was published in the EU Official Gazette on 30 December 2006 and amended by the *corrigendum* published on 18 January 2007. Additional amendments and integrations of this Regulation have been adopted with Regulations 107 and 109 of 2008 and 116 of 2010 and Article 49 of Regulation (EU) 1169/2011.

Moreover, Regulation (EC) 353/2008 defined the scientific criteria to support the proposed claims, and the guidance on the implementation of Regulation (EC) 1924/2006 on nutrition and health claims made on foods was adopted on 14 December 2007 (Standing Committee on the Food Chain and Animal Health, 2007).

A review of the situation in the European Union of nutrition and health claims and generic descriptors concerning foods and beverages has been developed by Silano (2014a).

According to Regulation (EC) 1924/2006, any non-mandatory message or image on the base of Community or national legislation, including figurative, graphical or symbolic representations stating, suggesting or implying, in any form, that a food has peculiar characteristics is defined as "claim".

"Nutrition claim" is any claim stating, suggesting or implying that a food may have particular beneficial nutritional properties, depending on:

a) the energy (caloric value) that:
   i) provides,
   ii) provides at a reduced or increased extent, or
   iii) does not provide, and/or
b) the nutritional substances that:
   i) contains,
   ii) contains at a reduced or increased extent, or
   iii) does not contain.

"Health claim" is any claim that states, suggests or implies a relationship between a food category, a food or one of its components and health. There are several types of health claims:

- functional (Article 13(1) and 13(5));
- reduction of a risk of disease (Article 14A); and
- development and health of children (Article 14B).

Not only sentences but also images such as a hearth or an athlete able to evocate beneficial effects for health constitute health claims, generic or specific. Claims which attribute to any foodstuff the property of preventing, treating or curing a human disease or refer to such properties remain prohibited (Directive 2000/13/EC). Table 5.5 shows one nutrition claim and three health claims for the same nutrient (calcium).

Another voluntary information foreseen by Regulation (EC) 1924/2006 is the "generic descriptor", that is a *traditional* information used to indicate particular classes of food products that might have a possible effect on human health (e.g. "digestive"). When authorized, the generic descriptor must be used in the same words used for at least 20 years.

Nutrition and health claims may be used in the labelling, presentation and advertising of foods placed on the market in the Community only if they comply with the provisions of Regulation (EC) 1924/2006.

The use of nutrition and health claims shall not:

(a) be false, ambiguous or misleading;
(b) give rise to doubt about the safety and/or the nutritional adequacy of other foods;
(c) encourage or condone excessive consumption of a food;
(d) state, suggest or imply that a balanced and varied diet cannot provide appropriate quantities of nutrients, in general. Derogations in the case of nutrients for which sufficient quantities cannot be provided by a balanced and varied diet, including the conditions for their application, may be adopted in accordance with the procedure referred to in Article 25(2) of Regulation (EC) 1926/2006, taking into account the special conditions present in the Member States; and
(e) refer to changes in bodily functions, which could give rise to or exploit fear in the consumer, either textually or through pictorial, graphic or symbolic representations.

Table 5.5   Nutrition and Health Claims for Calcium

<Contains calcium> (nutritional)

<Calcium is necessary for the normal structure of the bone> (functional)

<Calcium reduces the risk of osteoporosis> (risk reduction)

<Calcium is beneficial for the growth of children> (children health and development)

<Athlete or heart> (generic or unspecified)

The use of nutrition and health claims shall be permitted only if the following conditions are fulfilled:

(a) the presence, absence or reduced content in a food or category of food of a nutrient or other substance in respect of which the claim is made has been shown to have a beneficial nutritional or physiological effect, as established by generally accepted scientific evidence;

(b) the nutrient or other substance for which the claim is made:
  - is contained in the final product in a significant quantity as defined in Community legislation or, where such rules do not exist, in a quantity that will produce the nutritional or physiological effect claimed as established by generally accepted scientific evidence; or
  - is not present or is present in a reduced quantity that will produce the nutritional or physiological effect claimed as established by generally accepted scientific evidence;

(c) where applicable, the nutrient or other substance for which the claim is made is in a form that is available to be used by the body;

(d) the quantity of the product that can reasonably be expected to be consumed provides a significant quantity of the nutrient or other substance to which the claim relates, as defined in Community legislation, or, where such rules do not exist, a significant quantity that will produce the nutritional or physiological effect claimed as established by generally accepted scientific evidence;

(e) compliance with the specific conditions set out in Chapter III or Chapter IV, as the case may be.

The use of nutrition and health claims shall be permitted only if the average consumer can be expected to understand the beneficial effects as expressed in the claim (IPSOS and London Economics Consortium, 2013).

Nutrition and health claims shall refer to the food ready for consumption.

Considering the positive image that is associated with specific foods by nutrition claims (and also by health claims), nutrition labelling is mandatory for all food products that benefit of a nutrition or health claim, with the only exception of non-pre-packaged foods including fresh foods. The Commission has been charged with establishing and maintaining a "Community Register" of nutrition and health claims made on food, hereinafter referred to as "the Register", that will include, according to Regulation1924/2006, the:

(a) nutrition claims and the conditions applying to them as set out in Annex of the Reg.1924/2006;
(b) restrictions adopted in accordance with Article 4(5);
(c) the authorized health claims and the conditions applying to them provided for in Articles 13(3) and (5), 14(1), 19(2), 21, 24(2) and 28(6) and the national measures referred to in Article 23(3); and
(d) a list of rejected health claims and the reasons for their rejection.

Health claims authorized on the basis of proprietary data shall be recorded in a separate Annex to the Register, together with the following information:

(i) the date the Commission authorized the health claim and the name of the original applicant to whom the authorization was granted;
(ii) the fact that the Commission authorized the health claim on the basis of proprietary data; and
(iii) the fact that the health claim is restricted for use unless a subsequent applicant obtains authorization for the claim without reference to the proprietary data of the original applicant.

The administrative procedures to authorize the different types of nutrition and health claims are described in the following sections of this chapter, but it is important to stress here that all the authorizations depend on the positive evaluation by EFSA. Therefore, it is not surprising that EFSA has devoted many efforts to describe in detail the scientific methodologies adopted for the evaluation of nutrition and health claims (Table 5.6).

Table 5.6 EFSA's Approach to the Claim Evaluation

EFSA guidance:
Preparation and presentation of applications (2007, rev. 2011)
General principles for the substantiation of claims (2009, 2010)
Scientific requirements for the substantiation of specific types of health claims (2010–2012)
EFSA dialogue with applicants before acceptance and during the evaluation; EFSA's response to comments after publication – Application Desk
Stakeholder meetings to discuss general principles and specific topics
Presentations at other conferences

5.3.1 Nutrition Claims

Principles of nutrition were reviewed by Silano and Silano in 1999.

Nutrition claims are permitted only if they are listed in the Annex of EC Regulation 1924/2006 and are in conformity with the conditions set out in this Regulation for a specific food product.

Amendments to the Annex can be adopted in accordance with the procedure referred to in Article 25(2) of Regulation (EC) 1924/2006 and, where appropriate, after consulting EFSA. Where appropriate, the Commission shall involve interested parties, in particular, food business operators and consumer groups, in order to evaluate the perception and understanding of the claims in question.

In the case of comparative claims, a comparison may be made only between foods of the same category, taking into consideration a range of foods of that category. The difference in the quantity of a nutrient and/or the energy value has to be stated, and the comparison should relate to the same quantity of food.

Comparative nutrition claims compare the composition of the food in question with a range of foods of the same category, which do not have a composition, which allows them to bear a claim, including foods of other brands. The difference in the amount of a nutrient and/or in the energy value of different foods may be expressed in percentage or absolute value and must be referred to the same amount of food. In some cases, food products may have to be qualified, e.g. butter and margarine. When the indication "light" or "reduced in energy" is used, the underlying characteristics must be clarified, e.g. "light–50% less sugars". When the nutrients are removed from the composition, it is possible to use the indication "light–without sugars". In this type of nutrition declarations, the comparison is possible only between foods of the same category, taking into account all the food products present. If a specific branded food product has a composition representative of the market, the name of the product can provide the reference for the comparison if it is followed by the term "light", e.g. "X light", when X is the standard product. The food products subject to comparison must belong to food groups with similar nutrient contents (e.g. the comparison between milk and butter should be avoided). Moreover, some food groups are too broad to be considered as a category (e.g. dairy products). Therefore only categories such as "milks" or" cheeses" can be utilized. Moreover, to not mislead the consumer, it should be avoided to use an indication that, although being equivalent to 30% of a specific nutrient, concerns a food category irrelevant for the consumption of that specific nutrient (e.g. the indication "reduced in fats" as applied to bread). Annex 16.2 of the present book, corresponding to the Annex to Regulation (EC) 1924/2006 that lists the nutrition claims and relevant conditions applicable to low energy, energy reduced, energy free, low fat, fatfree, low saturated fats, saturated fats free, low sugars, sugars free, with no added sugars, low sodium/salt, very low sodium/salt, sodium/salt-free, source of fibre, high fibre, source of protein, high protein, source of [name of vitamin/s] and/or [name of mineral/s], high [name of vitamin/s] and/or [name of mineral/s], contains [name of the nutrient or other substance], increased [name of the nutrient], reduced [name of the nutrient], light/lite and naturally/natural.

The first amendment of the Annex to Regulation (EC) 1924/2006 has dealt with the addition of the following claims:

- source of omega-3 fatty acids;
- high in omega-3 fatty acids;
- high in mono-unsaturated fatty acids;
- high in poly-unsaturated fatty acids; and
- high in unsaturated fatty acids.

From the mentioned Annex it appears clear that for nutrients whose assumption should be controlled (e.g. saturated fats) to have health benefits only claims indicating their low/reduced levels in specific foods can be used, whereas for nutrients for which the consumption is recommended (e.g. vitamins) only claims indicating their increased/high contents in specific foods can be used.

An amendment was requested, but not agreed, to make possible the use of specific claims on re-formulated food to reduce the content of specific ingredients such as salt, saturated fatty acids and sugar also in smaller amounts than those foreseen by the Annex to Regulation (EC) 1924/2006, with a mention of the extent of the reduction. Some clarification has been provided on the use of the term "light" that should be following the conditions for the term reduced with a clear explanation of the underlying reasoning (e.g. light–50% less fats) and "natural" restricted to the case the food, as present in nature, meets the criteria indicated in the Annex.

When appropriate, a sentence indicating that the salt content is exclusively due to the natural presence of sodium can be added near the nutrition declaration. In case the nutrition declaration concerns nutrients such as mono-unsaturated or poly-unsaturated fatty acids, polyols, starch, fibre or any vitamin or mineral (Point 1 of Part A of Annex XIII), present in a "significant quantity" (as defined in Point 2 of Part A of Annex XIII), the amount of the nutrient will be declared.

5.3.2 Health Claims

Basic criteria for the acceptance of health claims include the:

- characterization of the food and its constituents;
- beneficial role for human health;
- evidence of cause–effect relationship;
- knowledge of the food amount needed for the indicated effect;
- representativity of available data for the target population; and
- need to take into account the globality of available scientific data and to evaluate all available evidence.

Health functional claims describe or refer to the following elements:

- role of a nutritional substance or of any other type of substance useful for growth, development and other functions of the body; or
- psychological and behavioural functions.

They can be based on:

- scientific evidence generally accepted (Article 13.1);
- general scrutiny;
- new scientific data and/or proprietary data (Article 13.5); and
- case-by-case authorization.

5.3.2.1 General Function Health Claims Already in Use (Article 13.1)

The claims ex Article 13(1) are the general functional claims authorized with an *ad hoc* procedure among all those already in use at the time Regulation (EC) 1924/2006 entered in force. A highly complex procedure has been necessary to gather and evaluate a very large number of claims in use in all the languages of the EU (as many as 44,000). About two years of intensive work were necessary to reduce the list of claims to be evaluated to 4,185 claims in English submitted to EFSA between January and June 2009. As many as 2,678 claims were evaluated by EFSA between October 2009 and June 2011. The first group of Article 13(1) claims was only approved by the EC on 16 May 2012 with Regulation (EU) 432/2012. The long time for the completion of the procedure of Article 13(1) is due, among others, also to the deliberations of the EC depending on the opinions of the Standing Committee for Food Chain and Animal Health and to the scrutiny procedure (90 days) of the European Parliament and Council.

As a consequence of the adoption of this Regulation, we find in the Commission Register (http://ec.europa.eu/nuhclaims/), 497 ID for a total of 222 claims concerning:

- 15 vitamins and 14 minerals: 18 for zinc; 15 for vitamin C; 10 for magnesium and vitamin B6; 9 for riboflavin and 8 for calcium, copper, folate and vitamin B12, each;
- about 10 substances such as fibres, omega-3 fatty acids, fermented red rice, olive oil polyphenols and creatine;
- some food categories such as nuts, foods with reduced content of sodium and meal substitutes;
- several comparative or substitutive claims such as "food with low or reduced fatty acids content", "food with a reduced content of salt", "MUFA/PUFA as substitutes for saturated fatty acids" and "sugar substitutes".

Nutrition claims concerning low alcoholic levels or reduction of the alcoholic content or the energetic content are permitted in case of beverages containing more than 1.2% by volume of alcohol. Another important Regulation in this area is Regulation (EU) 536/2013 that authorized the following new health claims:

- **Docosahexaenoic acid (DHA)**: "DHA contributes to the maintenance of normal levels of triglycerides in blood".
- **Docosahexaenoic acid and eicosapentaenoic (DHA/EPA)**: "DHA and EPA contribute to the maintenance of a normal blood pressure".
- **Docosahexaenoic acid and eicosapentaenoic acid (DHA/EPA)**: "DHA and EPA contribute to the maintenance of normal levels of triglycerides in the blood".
- **Alpha-cyclodextrin**: "The consumption of alfa-cyclodextrin within a meal containing starch contributes to the reduction of the post-prandial increase of blood glucose".
- **Fructose**: "The consumption of foods containing fructose determines a smaller increase of blood glucose than foods containing sucrose or glucose".

The analysis of the results obtained in terms of claims so far authorized by applying the Article 13(1) procedure indicates that we are in the presence of a true and careful translation into a language, also accessible to people devoid of specific scientific competence, of notions which can be generally found in a handbook/treaty on human biochemistry and nutrition. This result is quite peculiar but also characterized by a considerable strength in terms of communication. In very simple terms, these claims describe the link existing between the ingestion of specific vitamin, mineral or other substances and the normal development of a biochemical or physiological function essential to life. Table 5.7 stresses with some examples the fact that although the claim is authorized only after a positive EFSA opinion, the final decision on the wording of the claim depends on the EC decision. In case of changes from the version approved by EFSA , motivations have to be provided.

It should be noted that a considerable number of health claims already in use (e.g. those on botanicals or on probiotics) have not received a positive evaluation by EFSA. In 2010, after the rejection by EFSA of about 500 applications on botanical health claims due to a non-adequate characterization of the botanical preparations or to the non-demonstrated correlations between the use in humans of the botanical preparation and the beneficial health effects asserted, it became clear that substantiating the approximately 2,000 claims remaining applications on botanicals required more stringent proof of efficacy than was required for traditional herbal medicinal products (THMPs).

EFSA put these claim evaluations "on hold" while it deliberated how to resolve this dilemma. The on-hold claims can still be used as the EU

Table 5.7 Some Claims That Received a Positive Evaluation from EFSA and Were Modified in Their Formulations by the European Commission

Food/ingredient	EFSA	European Commission
Meat substitutes to control body weight	Maintenance of body weight following a body weight loss	The substitution of meat in one meal per day in an energy-restricted diet contributes to the maintenance of body weight following a weight loss
Glucomannan	Body weight reduction	Body weight reduction in the context of an energy-restricted diet
Copper, manganese riboflavin, selenium, etc.	Protection of DNA, proteins and lipids from oxidative stress	Cell protection from oxidative stress
Chitosan	Maintenance of the normal concentration of blood cholesterol	Maintenance of normal haematic level of cholesterol

Member States have taken this matter into their hands. Another group of claims that has had a very difficult life so far is the group of claims on probiotic products. Initially, the issue depended on a general lack of characterization of most products that made impossible any evaluation of the data provided to support the link between the cause and the effect as represented by the claim. After a first stop, in May 2011, the Article 13(1) procedure has been reactivated to make it possible for the interested companies to provide, through their respective Member States, the additional information needed to re-evaluate about 263 claims ex Article 13(1) concerning microorganisms inadequately characterized. Moreover, a similar approach was undertaken to re-evaluate about 40 claims on other products not approved by EFSA due to limited evidence, although not completely absent.

Although many applications for health claims on probiotics have been submitted for evaluation to EFSA, no application has received a positive opinion. Therefore, no claims on probiotics are listed on the EU Register as authorized for use. The probiotic claims that have been fully evaluated and rejected are listed as non-authorized in the EU Register.

5.3.2.2 General Function Health Claims Based on New Data Subject to Protection (Article 13.5)

The procedure of Article 13(5) is for the same kind of claims as that of Article 13(1), except for the fact that new data are necessary for the authorizations for which data protection may be requested. Therefore, these are health claims based on newly developed scientific evidence and/or applications which include a request for the protection of proprietary data. The first claim of this group authorized and included in the EU Register dealt with "tomato concentrate and piastrinic aggregation". An FBO wishing to add a health claim, based on newly developed scientific evidence and/or which includes a request for the protection of proprietary data to the Community list of permitted claims, has to apply for the inclusion of the claim in that list. Article 13(5) claims are authorized under the procedures detailed in Article 18 of Regulation (EC) 1924/2006, which is applicable since 1 February 2008. The application to be presented by the FBO to a competent authority in the Member State must include the following information:

- the name and address of the applicant;
- the nutrient or other substance, or the food or the category of food, in respect of which the health claim is to be made and its particular characteristics;

- a copy of the studies, including, where available, independent, peer-reviewed studies, which have been carried out with regard to the health claim and any other material which is available to demonstrate that the health claim complies with the criteria provided for in Regulation (EU) 1924/2006 on nutrition and health claims;
- where appropriate, an indication of the information which should be regarded as proprietary accompanied by verifiable justification;
- a copy of other scientific studies which are relevant to that health claim;
- a proposal for the wording of the health claim for which authorization is sought, including, as the case may be, specific conditions for use;
- a summary of the application; and
- the reasons for the request.

The application and any information supplied by the FBO will be sent by the competent Authority in Member State without delay to EFSA for a scientific assessment, as well as to the Commission and the other Member States for information (see also the Table 5.8).

EFSA will issue its opinion within five months from the date of receipt of the request. This time limit may be extended by up to one month if EFSA requires supplementary information to be submitted by the applicant. If this supplementary information is sought by EFSA, it must be submitted by the applicant within 15 days from the date of receipt of EFSA's request.

In order to prepare its opinion, EFSA will verify that the:

- health claim is substantiated by scientific evidence; and
- wording of the health claim complies with the criteria laid down in the Regulation.

EFSA will forward its opinion to the Commission, the Member States and the applicant, including a report describing its assessment of the health claim and stating the reasons for its opinion, according to its guidance regarding the submission of claims under Article 13(5) and the information on which its opinion was based and will make its opinion public. The applicant or members of the public may send comments to the Commission within 30 days from such publication. According to Article 17 of Regulation (EC) 1924/2006, within two months from receiving the opinion of EFSA, the Commission shall submit to the Standing Committee on the Food Chain and Animal Health a draft decision on the lists of

Table 5.8 Data to Be Provided with Applications for Authorizations Ex Articles 13(5) and 14 Claims

Part 1 – Technical and administrative data

  1.1. Table of contents

  1.2. Form for the presentation of the application

  1.3. General information

  1.4. Description of the health claim

  1.5. Summary of the application

  1.6. References

Part 2 – Characteristics of the food/ingredient

  2.1. Food ingredient

  2.2. Type or category of food

  2.3. References

Part 3 – Global summary of relevant scientific data

  3.1. Tabular summary of all the relevant studies identified

  3.2. Tabular summary of the data derived from relevant studies on human beings

  3.3. Written summary of the data derived from relevant studies on human beings

  3.4. Written summary of the data derived from relevant studies not carried out on human beings

  3.5. General conclusions

Part 4 – Global summary of all relevant scientific data

  4.1. Tabular summary of all relevant studies carried out on human beings

  4.2. Tabular summary of data derived from relevant studies carried out on human beings

  4.3. Written summary of data derived from relevant studies carried out on human beings

  4.4. Written summary of data derived from relevant studies not carried out on human beings

  4.5. General conclusions

permitted health claims including details on the claims, taking into account the opinion of EFSA, any relevant provisions of Community law and other legitimate factors relevant to the matter under consideration. The Commission shall without delay inform the applicant of the decision taken and publish details of the decision in *Official Journal of the European*

Union. Authorized claims will be added to the list in the Community Register, together with any conditions for use. If the claim is rejected, it will be added to the list in the Community Register, together with the reasons for rejection. For the procedure concerning the exclusive authorization of health claims motivated by "industrial property" of essential data, see Section 5.1.2.3.

5.3.2.3 Health Claims Ex Article 14A (Reduction of Disease Risk Claims) and 14B (Children Health and Development Claims)

In the initial phase, 360 health claims ex Article 14 were presented to EFSA, of which 128 have been approved and 146 withdrawn.

In Article 14A the risk reduction claims of a disease (e.g. "calcium/ vitamin D and reduction of the risk of osteoporotic fractures through a reduction of bone loss") are taken into account (Table 5.9). "Reduction of disease risk claim" means any health claim that states, suggests or implies that the consumption of a food category, a food or one of its constituents significantly reduces a risk factor in the development of a human disease.

Table 5.9 Some Article 14a Risk Reduction Claims

1. CHEWING GUM WITHOUT SUGAR and neutralization of the plaque acids, which are a risk factor for dental caries.

2. CHEWING GUM WITHOUT SUGAR and help in reducing the dental demineralization that is a risk.

3. OAT BETA-GLUCAN and reduction of blood cholesterol. The high cholesterol levels are a risk factor for coronary cardiac diseases.

4. VEGETABLE STEROL AND STANOL ESTERS ADDED TO A VARIETY OF FOODS SUCH AS SPREADABLE MARGARINE, MAYONNAISE AND DRESSINGS FOR SALADS AND MILK-BASED PRODUCTS SUCH AS YOGHURT AND CHEESES and lowering/reduction of blood cholesterol. A high level of blood cholesterol is a risk factor for the development of coronary cardiac diseases.

5. VEGETABLE STANOL ESTERS and reduction of blood cholesterol. A high level of blood cholesterol is a risk factor for the development of coronary cardiac diseases.

6. VEGETABLE STEROLS, PRESENT IN FREE FORM OR ESTERIFIED WITH FOOD FATTY ACIDS, and lowering/reduction of blood cholesterol. A high level of blood cholesterol is a risk factor for the development of coronary cardiac diseases.

In Article 14B claims concerning the health and development of children are considered (e.g. "iodine and normal growth of children") (Table 5.10).

The term "children", which is not defined in the Regulation, should be understood as reaching the end of the growth period. An indicative age limit of 18 years can be mentioned, but this indication does not intend to define "children" in the frame of this Regulation. Under Regulation (EC) 1924/2006, reduction of disease risk claims and claims referring to children's development and health may be made where they have been authorized in accordance with the procedure laid down in Articles 15, 16, 17 and 19 of the cited Regulation, which is identical to that described

Table 5.10 Some Article 14b Children Health and Development Claims

1. DOCOSAHEXAENOIC ACID (DHA) and the maternal assumption that contributes to the normal cerebral development of the foetus and breast-fed infant.

2. DOCOSAHEXAENOIC ACID (DHA) and contribution to the normal development of vision in children *until 12 months of age.*

3. DOCOSAHEXAENOIC ACID (DHA) and the maternal assumption that contributes to the normal development of the eye of the foetus and of the breast-fed infant.

4. ESSENTIAL FATTY ACIDS: ALPHA-LINOLENIC (ALA) AND LINOLEIC ACID (LA) and need for the normal growth and development of the children.

5. CALCIUM/VITAMIN D and need for the normal growth and development of children's bones.

6. CALCIUM and need for the normal growth and development of children's bones.

7. VITAMIN D and need for the normal growth and development of children's bones.

8. PHOSPHORUS and need for the normal growth and development of children's bones.

9. IODINE and contribution to the normal growth of children.

10. IRON and contribution to the normal cognitive development of children.

11. PROTEINS and need for the normal growth and development of children's bones.

12. CHEWING GUM SWEETENED WITH 100% XYLITOL and reduction of dental plaque development. An elevated level/content of dental plaque is a risk factor for the development of children caries.

in Section 5.1.2.2. Once authorized, they will also be included in a Community list of permitted claims, together with all the necessary conditions for their use. Regulation (EU) 1169/2011 that prevents labelling from attributing to any foodstuff the property of preventing, treating or curing a human disease, or referring to such properties, also applies. In the evaluation of health claims it is also very important to consider the detailed characterization of the food and its ingredients, as this makes it possible later on to check that the product present on the market is the one effectively authorized to the benefit of the claim. Other essential steps of the evaluation of health claims include the identification of:

- the cause–effect relationship between the food and the benefit for human health;
- the amount of food needed to obtain the benefit; and
- the representativity of the target population.

It is also necessary to show that all available data have been examined and not only the positive ones. In case the essential data to support the application are protected by "industrial property" and have not been published, it is possible to obtain the exclusive use of the claim for five years. In such a case it is necessary that the EFSA opinion and the EU authorization make an explicit reference to the property of the essential data.

According to the "Guidance on the Implementation of Regulation (EC) 1924/2006 on Nutrition and Health Claims Made on Foods", the following claims should be considered as Article 14 claims:

- health claims solely referring to the development and health of children and where the scientific substantiation is valid only for children. In this case, the scientific substantiation consists of data obtained on studies conducted with children. Example: "calcium is good for children's growth"; and
- health claims used on products intended exclusively for children, like follow on formulae, processed cereal-based foods and baby foods, as defined by Directive 2006/141/EC and Directive 2006/125/EC.

Some difficulties have been met in the application of the 14B claims to foods covered by Regulations on breast-fed infants and follow-on formula (Directive 92/52/CEE and 2006/141/CE) and on cereal-based foods and other foods intended for breast-fed infants and children (Directive 2006/125/CE). As these Regulations define in great details the compositions of the above-mentioned food products to ensure their efficacy, some doubts have emerged about accepting the use of specific health claims on particular non-regulated components instead of updating the above-mentioned Regulations by adding the specific components considered to deserve a health claim. One of the more problematic aspects that emerged with respect to the authorization of Article 14 claims has been the high costs necessary for the supporting data needed. In fact, the administrative cost to obtain an authorization ex Article 14 amounts to some thousand euros, but a randomized prospective case–control study with placebo on human health volunteers to support an application for a health claim may cost more than 250,000 Euros.

5.3.3 Nutritional Profiles

According to Article 4 of Regulation (EC) 1924/2006, the EC should have adopted, according to the procedure of Article 25, Paragraph 2, within 19 January 2009, specific "nutritional profiles" applicable to foods or some categories of food to be able to bring nutrition or health claims, as well as conditions concerning the use of nutritional and health claims in relation to nutritional profiles.

Nutritional profiles should have been defined by taking into account:

a) the amounts of specific nutritional or other constituents of the food product, e.g. saturated fats, saturated fatty acids, transfatty acids, sugars and salt/sodium;
b) the role, importance and contribution of the food or food category in the diet of the general population or, when appropriate, of specific population groups at risks;
c) the global nutritional composition of the specific food as well as the presence of nutritional substances with recognized health effects.

When determining nutritional profiles, the EC should undertake consultations with stakeholders.

The EFSA's opinion on nutritional profiles has been adopted on 31 January 2008 with the main aim of preventing the use of nutritional profiles for food categories with high contents of salt, sugars and/or saturated fats. Moreover, EFSA compiled, in collaboration with the Member States, a database to test different profile scenarios.

In the frame of the discussion of the new Regulation on labelling, the European Parliament discussed, without reaching a consensus, the abrogation of Article 4 of Regulation (EC) 1924/2006, and the adoption process of nutritional profiles is still pending.

6

Food and Feed Hygiene, Official Controls and RASFF, Agri-Food Frauds and Unfair Commercial Practices

6.1 INTRODUCTION

The detailed implementation of the principles, procedures, organization and scientific bases introduced by Regulation (EC) 178/2002 has been made possible from a very intensive regulatory production in many different sectors that has considerably changed the large number of Regulations adopted before 2002 (Silano,1994; Silano and Silano,1997).

Since 1 January 2006, the food business operators and the responsible authorities for official controls have to respect and apply the rules of the so-called hygiene package intended to ensure the safety of food manufacturing process and of relative control activities. This package includes the following Regulations:

- Regulation (EC) 852/2004 on the hygiene of food products in general;
- Regulation (EC) 853/2004 that establishes specific norms for food products of animal origin;
- Directive 2004/41/EC concerning the hygiene of food products and sanitary measures for the manufacturing and marketing

69

of specific food products of animal origin intended for human consumption;

- Regulation (EC) 882/2004 on the official controls of foods and feeds, that, since 14 December 2019, has been repealed by Regulation (EU) 2017/625 on official controls and other official activities performed to ensure the application of food and feed law, rules on animal health and welfare, plant health and plant protection products. Regulation (EU) 2017/625 applies in general since 14 December 2019. However, requirements in relation to Reference Laboratories and Reference Centres, specified in Articles 92-101, have applied since 29 April 2018 instead of Articles 32 and 33 of Regulation (EC) 882/2004; and
- Regulation (EC) 854/2004 on the specific control measures for the products of animal origin intended for human consumption.

6.2 FOOD AND FEED HYGIENE: HACCP AND SELF-CONTROL (REGULATION (EC) 852/2004)

General rules on the hygiene of foodstuffs have been adopted with Regulation (EC) 852/2004. The key elements of these Regulations are:

- the implementation of the "Hazard Analysis and Critical Control Points" (energy/calories, medium acidity and "free water" activity) approach;
- the manuals of good hygiene practices in specific sectors;
- the standard operation procedures;
- the registration or approval for certain food establishments; and
- specific requirements to be ensured in the construction, maintenance and management of food manufacturing plants, including personnel hygiene and temperature control.

The "Hazard Analysis and Critical Control Points" approach to be carried out by the food business operator consists in the identification and management of risk factors through critical control points depending on each food chain.

Regulation (EC) 852/2004 clearly states that each business operator of the food sector must ensure the safety of the products from fields to table as well as the principle that the primary responsibility of food safety belongs to the operators of this sector (Article 1, Part 1, letters a) and b) of Regulation (EC) 852/2004).

Food business operators (FBOs) have the duty to ensure that all the different steps of manufacturing, transforming and distributing the food products which are under their control are carried out in compliance with the hygiene criteria and requirements established by Regulation (EC) 852/2004. Obviously, the food business operators have also to comply with the hygiene requirements of other Regulations such as the prescriptions on food products of animal origin or the prescriptions concerning maximum levels of intentionally added and unintentionally added substances (see Chapter 7).

Regulation (EC) 852/2004 identifies a number of different tools to ensure the compliance of food business operators with the rules of hygiene, including the procedures based on the HACCP system and the implementation of general hygiene rules listed in this Regulation. Annex 16.3 of the present book deals with primary production requirements. Annex 16.4 deals with the general hygiene requirements applicable to:

- all food premises, except movable and/or temporary premises (such as marquees, market stalls, mobile sales vehicles), where foods are regularly prepared for placing on the market and vending machines;
- all rooms where food is prepared, treated or processed, except dining areas and movable and/or temporary premises (such as marquees, market stalls, mobile sales vehicles), premises used primarily as a private dwelling-house but where foods are regularly prepared for placing on the market and vending machines;
- movable and/or temporary premises (such as marquees, market stalls, mobile sales vehicles), premises used primarily as a private dwelling-house but where foods are regularly prepared for placing on the market and vending machines;
- all transportations; and
- all stages of production, processing and distribution of food.

The FBOs have to elaborate, implement and maintain one or more permanent procedures of corporate self-control based on the principles of the risk analysis system and control critical points (HACCP, Article 5, Part 1 of Regulation (EC) 852/2004). Such an obligation applies only to the food business operators who are active in the food manufacturing, transformation and distribution steps subsequent to the primary production and associated activities (Article 5, Part 3 of Regulation (EC) 852/2004).

The critical control points depend on the risk factors to be controlled during the relevant industrial step and include:

- energy/calories provided or subtracted;
- medium acidity (pH= -log [H+]); and
- free water activity (*aw*).

For instance, in the industrial milk pasteurization the critical control point is the energy transferred to milk through maintenance of the milk for specific times at the predetermined temperature of 80°C. In the case of vegetable preserves the critical control point is the medium acidity (pH = 4.5), whereas for the processing of raw ham the critical control point is free water activity. It is fundamental that in each industrial procedure the critical control point is automatically self-monitored and that the relative registrations are kept to be shown later on to food inspectors.

To ensure full implementation of hygiene and safety prescriptions, food business operators must also implement the many specific measures which integrate the HACCP system. This is the case of measures intended to comply with:

- the requirements of temperature control in different compartments of the food plants and the maintenance of the cold chain for foods that cannot be stored in the condition of safety at room temperature;
- the need to check the compliance of food products with the relevant microbiological criteria; and
- the overall objectives of Regulation (EC) 852/2004.

Other tools which can be used by the food business operators include:

- the use of manuals of correct hygiene practice that facilitate the compliance with hygiene rules; and
- the elaboration of corporate hygiene plans that greatly facilitate the compliance with hygiene and safety rules.

It should also be mentioned that in Regulation (EC) 852/2004 (Article 1, Part 1, letter (g) and Article 10), it is clearly stated the principle that imported foods must have been produced under identical or equivalent hygiene standards than those manufactured within the country. Regulation (EC) 852/2004 and Regulation (EC) 853/2004 both consider the institution of a system of registration and recognition of all food establishments by the competent authority of each Member State to make possible the implementation of the official controls.

According to Regulation EC 852/2004 each establishment manufacturing, transforming and distributing food products has to notify its activity in order to be registered. An authorization in response to this notification is only due if required by national laws or by Regulation (EC) 853/2004.

6.3 SPECIFIC HYGIENE RULES FOR FOODS OF ANIMAL ORIGIN (REGULATION (EC) 853/2004)

In addition to the general hygiene requirements applicable to all food categories, there are specific hygiene prescriptions relevant for food categories of animal origin as they may present particular microbiological and chemical risks for human health and, therefore, need the implementation of specific hygiene rules.

A first issue to be mentioned is the fact that products of animal origin should only be obtained from healthy animals characterized by health police indicated by the pertinent Regulation and, especially, by Article 3, Part 3 of Directive 2002/99/CE. Moreover, this Directive also defines the health criteria to be satisfied by-products of animal origin imported from third countries.

The requirements common to several products of animal origin are detailed in Annex II of Regulation (EC) 853/2004, whereas specific hygiene requirements for particular products of animal origin, including (i) meat of domestic ungulates; (ii) meat from poultry and lagomorphs; (iii) wild game meat; (iv) minced meat, meat preparations and mechanically separated meat; (v) meat products; (vi) live bivalve molluscs; (vii) fishery products; (viii) raw milk and dairy products; (ix) eggs and egg products; (x) frogs legs and snail; and (xi) other materials, such as gelatin and collagen, are specified in Annex III of Regulation (EC) 853/2004.

The establishment working products of animal origin and for which specific requirements are identified by Annex III of Regulation (EC) 853/2004 must be clearly recognized (Article 4, Part 2 of Regulation (EC) 853/2004). In addition to slaughterhouses and slicing labs, they include establishments for:

- ground meats and mechanically separated meats;
- meat products;
- depuration and shipping of live bivalve molluscs;
- fishing products;
- milk and derivatives; and
- eggs and egg products.

An exclusion from the recognition obligation exists for the establishments only involved in:

- primary production;
- transport; and
- storage of products not requiring controlled thermal installations.

In addition to complying with the general rules of Regulation (EC) 178/2002 (and in particular with the requirement of traceability – Article 18), the business operators of foods of animal origin must also comply with the requirements of health and identification marking (Article 5 of Regulation (EC) 853/2004).

Each stock of products of animal origin must be accompanied with adequate certificates or documents if this is requested by the Regulation and, in particular, by its Annexes II and III (Article 7 of Regulation (EC) 853/2004).

To eliminate the superficial contamination of products of animal origin, food business operators can use only drinking water or, if allowed, clean water (Article 3, Part 2 of Regulation (EC) 853/2004). Other substances can be utilized only if they have been formally approved at a European level for that use on the basis of Regulation (EC) 853/2004.

Decontamination substances are applied to foods of animal origin intended for human consumption to remove microorganisms that can cause diseases, such as *Salmonella* and *Campylobacter*, from their surface or toxic chemical substances (e.g. see the opinion on organic lactic and acetic acids to reduce microbiological surface contamination on pork carcasses and pork cuts that was adopted by the EFSA CEP Panel, in collaboration with the BIOHAZ Panel, in December 2018).

To ensure a high level of human health protection, food products imported in the EU must also comply with specific hygiene requirements. In fact, the food business operators may import products of animal origin from third countries only if the requirements indicated in Article 6 of Regulation (EC) 853/2004 are ensured.

6.4 OFFICIAL FOOD CONTROLS AND THE RAPID ALERT SYSTEM FOR FOOD AND FEED

6.4.1 Legislation on Official Food Controls

The Commission website, the EU official control rules provide national enforcers and the European Commission with the necessary powers to ensure the effective enforcement of regulatory requirements and with mechanisms that allow the full cooperation of all parties involved in ensuring the correct application of the law across national borders. The Official Controls Regulation also provides the European Commission with audit and control powers in the EU countries and third countries, and with the power to take action at EU level. According to the European Commission website, Regulation (EC) 882/2004 has aimed at creating an integrated and uniform approach to official controls along the agri-food chain throughout the Union. Its purpose is to allow competent authorities in the EU countries to verify compliance with food and feed laws and to:

- prevent or eliminate risks which may arise, either directly or via the environment, for human beings and animals, or reduce these risks to an acceptable level;
- guarantee fair practices as regards trade in food and feed; and
- ensure the protection of consumers' interests, including labelling of food and feed and any other form of information intended for consumers.

According to the EC website, key elements of Regulation (EC) 882/2004 on official controls include:

- official controls should be carried out regularly, on a risk basis and with appropriate frequency;
- official controls should be carried out at all stages of the food chain on domestic produce, as well as on imports and exports;
- competent authorities may delegate specific tasks to official control bodies, under certain conditions;
- specific rules are laid down for official controls carried out on imported products;
- regular training for competent authority staff is an obligation for the EU countries;
- training programmes for competent authorities' staff are funded by the European Union (Better Training for Safer Food see Chapter 13);
- framework for the designation of EU Reference Laboratorises;
- rules on the design and implementation of multi-annual national control plans prepared by the Member States to ensure the effective implementation of the Regulation; and
- possibility of coordinated control plans on an *ad hoc* basis.

Regulation (EC) 882/2004 has been reviewed and after a transition period, replaced by the Official Controls Regulation (EU) 2017/625 adopted by the European Parliament and Council on 5 March 2017, published in *Official*

Journal of the European Union on 7 April 2017 and entered into force on 27 April 2017. Regulation (EU) 2017/625 addresses official controls and other official activities performed to ensure the application of food and feed law, rules on animal health and welfare, plant health and plant protection products. The new rules are intended not only to repeal and replace Regulation (EC) 882/2004 but also to:

- repeal: Regulations (EC) No. 854/2004 and Council Directives 89/608/EEC, 89/662/EEC, 90/425/EEC, 91/496/EEC, 96/23/EC, 96/93/EC, 97/78/EC and Council Decision 92/438/EEC; and
- amend with respect of control rules: Regulations 999/2001/ EC, 396/2005/EC, 1069/2009/EC, 1107/2009/EC, 1151/2012/ EU, 652/2014/EU, 2016/429/EU, 2016/2031/EU of the European Parliament and the Council; Council Regulations 1/2005 and 1099/2009; and Council Directives 1998/58/EC, 1999/74/EC, 2007/43/EC, 2008/119/EC and 2008/120/EC.

The new rules have become gradually applicable, with the main application date being 14 December 2019, together with a number of Official Controls Regulation adopted as delegated and implementing acts (see Table 6.1).

The main elements and innovations of Regulation (EU) 2017/625 are listed in Table 6.2.

6.4.2 Official Controls on Products Coming from the Internal European Union Market

The official controls of food and beverages, generally carried out by the competent authorities in Member Countries, consist of:

  (i) Administrative checks;
 (ii) Collecting representative samples during food and feed manufacturing to check the quality of the manufacturing processes and collecting representative samples from food and feed products ready to enter or that have entered the food chain;
(iii) Preparing extracts from the collected samples and performing analyses for suspect or non-target screening; and
(iv) Formal checks on labelling and a number of different documents accompanying foods and feeds.

Table 6.1 Delegated and Implementing Acts of Regulation (EU) 2017/626 on Official Controls

- Delegated Regulation (EU) 2019/624 concerning official controls of the production of meat and for production areas of live bivalve molluscs
- Del. Regulation (EU) 2019/2090 concerning requirements for official controls on residues of pharmacologically active substances authorized in veterinary medicinal products or as feed additives
- Del. Regulation (EU) 2019/478 concerning subjects composite products, hay and straw to systematic border control
- Del. Regulation (EU) 2019/2122 concerning categories of animals and goods which are exempted from Border Control Post controls
- Del. Regulation (EU) 2019/1081 concerning training courses requirement for BCP staff
- Del. Regulation (EU) 2019/1602 concerning CHED accompanying consignments to their place of destination
- Del. Regulation (EU) 2019/2124 concerning transit, shipment and transportation through the Union
- Del. Regulation (EU) 2019/2123 concerning the identity and physical checks at control points. Documentary checks at a distance from BCP
- Del. Regulation (EU) 2019/1012 concerning re-designation of BCP and derogations from BCP requirements
- Del. Regulation (EU) 2019/2074 concerning specific controls for animals and goods returning to the Union after the refusal of entry by a third country
- Del. Regulation (EU) 2019/1666 concerning monitoring of arrival of consignments at the place of destination of products of animal origin and animal by-products
- Del. Regulation (EU) 2019/2125 concerning specific official controls for wood packaging materials
- Del. Regulation (EU) 2019/2126 concerning specific controls for categories of animals and goods which are exempted from BCP controls
- Del. Regulation (EU) 2018/631 concerning establishing plant EURLS
- Del. Regulation (EU) 2019/625 concerning requirements for the entry into the Union of certain animals and goods intended for human consumption + lists of third countries authorized to export to the EU
- Del. Regulation (EU) 2019/2127 concerning the change of the date of application of certain provisions of Directives 91496/EEC, 97/78/EC and 2000/29/EC
- Implementing Regulation (EU) 2019/627 concerning rules of official controls on the production of meat and for the production of live bivalve molluscs
- Implementing Regulation (EU) 2019/66 concerning official controls on plants, plant products and the objects in order to verify compliance with Union plant health rules

(Continued)

Table 6.1 (Continued)   Delegated and Implementing Acts of Regulation (EU)
2017/626 on Official Controls

- Implementing Regulation (EU) 2019/2007 concerning the identification of animals and goods subject to systematic border controls with CN codes
- Implementing Regulation (EU) 2019/1793 concerning: (i) lists of goods(feed and food of non-animal origin) from certain third countries subject to a temporary increase of official controls at BCPs (ii) consolidation of existing emergency measures establishing special import conditions for feed and food of non-animal origin from a certain third country; and (iii) establishing rules on the frequency of identity and physical checks, including laboratory tests
- Implementing Regulation (EU) 2019/2130 concerning detailed rules on the operations to be carried out during and after BCP checks
- Implementing Regulation (EU) 2019/2129 concerning rules for the uniform application of the appropriate frequency rate of identity and physical checks on consignments of animals and goods intended for release for free circulation
- Implementing Regulation (EU) 2019/1013 concerning minimum time requirements for prior notification before the physical arrival of consignments
- Implementing Regulation (EU) 2019/1014 concerning BCP facilities and listing requirements
- Implementing Regulation (EU) 2019/1873 concerning coordinated intensified controls at PCBs
- Implementation Regulation (EU) 2019/2128 concerning establishing the model of certificate for goods which are delivered to vessels leaving the Union and intended for ship supply or NATO or US military base
- Implementation Regulation (EU) 2019/530 concerning EURLs for pests of plants on insects and mites, nematodes, bacteria, fungi and oomycetes, viruses, viroids and phytoplasmas
- Implementation Regulation (EU) 2019/1685 concerning EU Reference Centres for animal welfare (poultry and other small animals)
- Implementation Regulation (EU) 2018/329 concerning European Union Reference Centre for animal welfare (pigs)
- Implementation Regulation (EU) 2019/723 concerning standards model form for MS annual reports
- Implementation Regulation (EU) 2019/626 concerning requirements for the entry into the Union of certain animals and goods intended for human consumption + lists of third countries authorized to export to the EU
- Implementation Regulation (EU) 2019/628 concerning official model certificates for certain animals and goods
- Implementation Regulation (EU) 2019/1715 concerning IMSOC, including the exchange of information between authorities and the format of the CHED

Source: European Commission Website.

Table 6.2   Main Elements and Innovations in Regulation (EU) 2017/625 on Official Controls

According to the European Commission, the main elements and innovations of Regulation (EU) 2017/625 include:

Risk-based approach

Official Controls on the Operator's Process and Activities

Transparency of official controls – Greater Accountability of the Competent Authorities

Delegation of Control Tasks of the Competent Authorities

Operators' Obligations

Official Laboratories

Reference Laboratories and Centres

Sampling, Analysis, Test and Diagnosis

Border Controls

Official Certification

Administrative Assistance and Cooperation

Financial of Official Control and Other Official Activities

Enforcement Action by Competent Authorities

Enforcement Measures

Penalties

Sector Specific Control Rules

Relation between Official Control Regulations and the Animal Health Law and the Plant Health Law

Commission Empowerments: Future-Proofs Legislations

Source: European Commission Website.

The organization of the system of official controls depends considerably on the Member State. Figures 6.1 and 6.2 show the current organization in Italy of the system of official controls.

In Italy, for the specific establishments requiring authorization before the start of the activity, the local health units (Aziende Sanitarie Locali) may carry out two types of inspections: those finalized to the authorization release and those intended to verify the good functioning of all the activities at different times. The inspections to verify the good functioning at different times apply to all the food-related establishments.

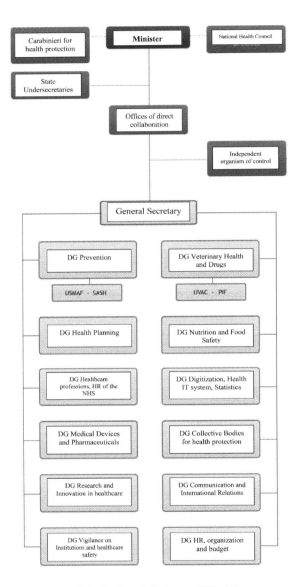

Figure 6.1 Organigram of the Italian Ministry of Health.

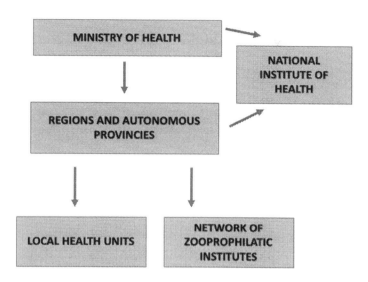

Figure 6.2 Framework of the Italian system for official control.

The Italian local health units are also in charge of issuing the administrative sanctions. In case the violations also represent offences to consumer safety, they should be reported to the competent judge for possible penal sanctions. In Italy, the official controls are carried out following the recommendations of the European Commission, Ministry of Health, Regions and Local Health Units for regularly systematic controls or in cases of food-related intoxications or specific crises (Figure 6.2). The food samples are, in general, sent for the analytical chemical or biological or physical controls to the laboratories of Zooprofilactic Institutes and in some cases to the Environmental Agencies. The analyses of revision to confirm the results of the first-instance analyses are carried out in Italy by the Istituto Superiore di Sanità (National Institute of Health) in Rome. Revision analyses are not carried out in case the presence of a representative of the interested food business operator has been ensured at the first-instance analysis (as it is generally the case for perishable food samples). Currently, in Italy the Regulation on the revision of analysis is being revised. Very important are obviously also the official controls carried out on voluntary and mandatory labelling and other information provided on food products. In Italy, these controls are carried out under the prescriptions of the legislative decree n. 27 of 7 February 2017 and n. 231 of 15 December 2017 concerning, respectively, the sanctioning disciplines of violations of Regulation (EC) 1924/2006 and Regulation (EU) 1169/2011. The main objectives of food label official controls are to check that the FBO has adequate documentations, including laboratory analyses, to support the information provided in the label.

For the official controls of food information, there are in Italy competences also for Institutions other than those of the National Health Service such as the Ministry of Agriculture and Forests, especially for foods with denominations of protected geographical origins and other claims such as those for biological foods. Other authorities being competent for the official controls on information provided to consumers include in Italy the Competition and Market Authority (Consumption Codex – Legislative Decree 206/2005). This authority has investigative powers, the possibility of temporarily suspending incorrect commercial practices and may also impose substantial financial sanctions. The Code of Self-Discipline is also able in Italy to modify or stop some unfair commercial practices.

6.4.3 Official Controls on Imported Products in the EU Market

Strict import rules with respect to food and feed hygiene, consumer safety and animal health status in the European Union, which is a major importer of food and feed, aim at assuring that all imports fulfil the same high standards as products from the EU itself. Import controls are crucial in verifying compliance of food and feed products with relevant requirements. The current approach to import controls varies according to the sector. Mandatory channelling of products to border control entities and uniform frequencies for checks applies to live animals, products of animal origin, plants and plant products because of the risk those commodities might pose in relation to animal or plant health, respectively. The vast majority of other products of relevance for the food chain are not channelled through specific border entities and do not need to undergo mandatory checks prior to their entry into the EU. It concerns for example feed and food which are of non-animal origin – including certain composite products, additives and other substances that might impact the characteristics of food and feed or materials intended to enter into contact with food. One notable exception may be the case of food and feed of non-animal origin which has been temporarily subjected to mandatory border controls due to the existence of an identified risk (i.e. the products listed under Regulation (EC) 669/2009). In relation to imported food products, the official controls are carried out in Italy by the organs of the Ministry of Health, denominated Veterinary Offices for Community

Obligations (UVAC), responsible for assisting Community Veterinary and Zootechnical Authorities and are present throughout the all Italian territory with their 17 offices, and the 38 border Inspection posts (BFC) competent for border control on animal health. These offices are also responsible for adopting restrictive measures on importation when needed.

6.4.4 The Rapid Alert System for Food and Feed

The RASFF was put in place in 1979 to provide food and feed control authorities with an effective tool to exchange information about measures taken responding to serious risks detected in relation to food or feed. This helps the Member States to act more rapidly and in a coordinated manner in response to a health threat caused by food or feed. A considerable evolution of the RASFF has taken place thanks to the availability of new technological instruments and to the practical experience gained by dealing with a number of emergencies (for example, from the initial fax transmissions of alerts, the system has moved to the e-mail transmissions and, then, to the more recent RASFF web).

The RASFF network includes all the EU Member States, EEA countries (Norway, Liechtenstein and Iceland), the EFTA Secretariat coordinating the input from the EEA countries, the EFSA and the EC as the manager of the system. Switzerland is a partial member of the system. Although a product which is non-compliant with food safety regulations must be withdrawn from the market irrespective of health risks, the RASFF way of working is risk-based, focusing on alert notifications for which a rapid action is required from identified members of the network. The need for rapid action is decided by the level of risk. Therefore, the exceedance of a legal limit, i.e. "no compliance" with a regulatory framework (European or National), does not systematically trigger a RASFF notification. An evaluation of the level of risk is necessary to decide that an alert notification should be issued. EFSA has been requested by the European Commission (Article 31 mandate) to propose a risk evaluation methodology that would allow a rapid and consistent risk-based classification of RASFF notification (three work packages: toxicity, exposure and IT tool). In March 2019, the risk evaluation tool entitled "Risk Evaluation of Chemical Contaminants in Food in the Context of RASFF Notifications. Rapid Assessment of Contaminants Exposure (RACE) has been developed for the evaluation of contaminants (arising from food contact materials, pharmacologically active substances and other food contaminants) in food. The risk evaluation has been based on the assessment of toxicological properties and dietary exposure of the agent involved in the alert notification. The result, expressed as the comparison of exposure to a relevant toxicological reference point, may lead to the classification of a specific alert as "no risk"; "low probability of adverse health effects" or "low concern for public health"; "potential risk"; or, simply, "risk". The above-methodology, developed by an EFSA *ad hoc* working group, provides guidance on a rapid assessment, based on (i) RASFF SOPs; (ii) EU legislation; (iii) extractions from the RASFF database; and (iv) EFSA scientific publications on:

- risk assessment of contaminants in food and feed;
- harmonized approach for the assessment of substances which are genotoxic and carcinogenic;
- the margin of exposure approach (MoE);
- the threshold of toxicological concern (TTC); and
- reference points for action (RPAs).

To facilitate the evaluation of exposure, the "Rapid Assessment of Contaminants Exposure (RACE) Tool" was developed by using the food consumption information from the EFSA Comprehensive European Food Consumption Database to provide estimates of acute and chronic exposure from single food and compares the result to the relevant toxicological reference points. This methodology provides a transparent set of criteria that can support the decision to notify the alert in RASFF and increase its transparency and harmonization by using EFSA/MS data on food consumption, standardized data dictionaries and open food tox. Some legal challenges associated with the international character of the RASFF system and with specific outcomes have been reviewed by Vincente Rodriguez Fuentes (2017).

6.5 ANALYSIS OF FLEXIBILITY PRINCIPLES IN FOOD SAFETY IN THE EU

The term "flexibility" is used in relation to the EU Hygiene Package to reflect the possibility of applying the rules in a way that is proportionate to the risk posed by particular food operations and establishments. Flexibility is, therefore, lined with the interpretation and implementation of certain hygiene provisions in specified areas and topics regarding special circumstances and structures of FBOs. "Flexibility" means to move the focus from requisites/stringent requirements to objectives. When terms such as "where necessary", "where appropriate", "adequate" and "sufficient" are used in the hygiene Regulations then it is up to the FBO to decide whether a requirement is necessary, appropriate, adequate or sufficient to achieve a food hygiene package (FHP) objective. Within the concept of flexibility, it is important that the FBO guarantees that the objectives of the FHP are met and that the safety of the food has not been compromised. A demonstration of the achievement of these objectives is thus essential and should be provided by the FBO in their Food Safety Management System (FSMS).

As highlighted by Mancuso et al. (2018), the FHP offers three different options for flexibility:

1. to exclude some activities from the FHP scope;
2. to grant adaptations of certain requirements laid down in the FHP:
   - to enable the continued use of traditional methods of production;
   - to accommodate the needs of food businesses situated in regions that are subject to specific geographic constraints; and
   - to adapt requirements on the construction, layout and equipment of establishments;
3. To grant derogation exemptions from certain requirements, laid down in the Annexes of the FHP, in particular for food with traditional characteristics (Article 7 of Regulation 2974/2005).

When making use of flexibility provisions, Member States have to adopt specific national measures. Therefore,

- based on the principle of subsidiarity Member States are best placed to find solutions to local situations; and
- based on the transparency principle, each draft of such national measures must be notified to the European Commission and to the other Member States.

The notification of national draft measures, in accordance with Directive 2015/1535, is managed through the Technical Regulation Information Service (TRIS) system, a software used by the Member States and the Commission to transmit the draft measures. Based on this database, an analysis has been possible of the notification criteria (see Mancuso et al., 2018). As many as 175 notifications were retrieved in this survey in the period 2004–2017, although the number of substances involved is higher due to the fact that some of the notifications are related to more than one subject.

The paper published by Mancuso et al. (2018) provides a rather detailed analysis of the flexibility notifications in the period 2004–2017 by the Member State, by year, by the type of foodstuffs, by quantity and by flexibility measures. This paper also explains how the different aspects of flexibility are applied in different countries. Moreover, the distribution of notifications over time and on the various areas subject to flexibility gives an indication about which aspects of food safety are most in need of an in-depth analysis from a regulatory point of view (e.g. small establishment and small quantities).

6.6 AGRI-FOOD FRAUDS: EU FOOD FRAUD NETWORK, ADMINISTRATIVE ASSISTANCE AND COOPERATION SYSTEM AND EU COORDINATED ACTIONS

Food fraud exists whenever individuals or businesses intentionally deceive the consumers, gaining an unfair advantage and violating the agri-food chain legislation. The infringements to the EU agri-food chain legislation may also constitute a risk to human, animal or plant health, to animal welfare or to the environment as regards genetically modified organisms (GMOs) and pesticides. Four key operative criteria are referred to for distinguishing whether a specific case should be considered as fraud or simply as non-compliance: if a case matches all four criteria, then it is considered a suspicion of fraud. These criteria correspond to the rules currently in place in the EU countries to report frauds:

1. **violations of EU law:** it involves violations of one or more rules codified in the EU agri-food chain legislation.
2. **intentions:** it can be verified through a number of factors which give strong grounds to believe that certain non-compliances are not happening by chance, such as the replacement of a high-quality ingredient with a lower quality one (if a contamination due to production processes is possible, when an ingredient is mostly replaced with a lower quality one there is a substitution, which often implies fraudulent intent).
3. **economic gains:** it implies some form of direct or indirect economic advantages.
4. **deception of customers:** it involves some form of deception of the customers/consumers (e.g. altered colours or altered labels, which mystify the true quality or, in worse cases even the nature of a product).

Fraudulent practices in the food sector may also lead to public health risks and damage. Some examples that have drawn worldwide attention are reported in the EC website for the years 1981, 1999, 2008 and 2013 and 2017. Cooperation between officials with EU agri-food chain knowledge, police and customs officers with investigative powers, judges and prosecutor administrations is very important at the national and EU level. Since 2013, the EU Food Fraud Network allows the Member States and some other countries in Europe to exchange information and to cooperate on a voluntary basis in cases of violations of the EU agri-food chain legislation of cross-border nature. This helps the European countries involved to work in accordance with the rules laid down in the Official Controls Regulation in matters where the national authorities are confronted with possible intentional violations of the EU agri-food chain legislation with a cross-border impact.

The EU Food Fraud Network is a cooperative approach based on trust. It connects the bodies designated by each Member State, some other European countries (Switzerland, Norway and Iceland), representatives from the European Commission services and representatives from Europol. It allows assisting and coordinating communication between competent authorities and, in particular, transmitting and receiving requests for assistance. The liaison bodies are required to exchange information necessary to enable the verification of compliance with EU agri-food chain legislation with their counterparts and, in certain cases, with the Commission, where the results of official controls require action in more than one country. As reported on the European Commission website, since 2015, members of the EU Food Fraud Network exchange information within the administrative assistance and cooperation (AAC) system. This is the information technology system developed and managed by the European Commission for EU Member States to exchange data in a structured manner regarding non-compliances and potential intentional violations of the EU agri-food chain legislation. Implementing Decision 2015/1918/EU details all the rules for the functioning of the administrative assistance and cooperation procedure. The AAC system consists of two parts:

- one part dedicated to the Food Fraud Network; and
- one part dedicated to every request for administrative assistance and cooperation which does not present profiles of human or animal risks, health risks and/or suspicion of potential intentional violations of the EU agri-food chain legislation.

The AAC system works in parallel with the RASFF, operational since 1979 (see also Section 6.2.4). It enables information to be shared efficiently and allows swift reaction when risks to public health are detected in the food chain. The AAC system is a tool for an EU country to rapidly contact the competent authorities of another EU country and share the information which can lead to administrative actions and/or sanctions or judicial proceedings. The AAC system is a "good tool" for the most efficient choice:

- for many intentional violations of the EU agri-food chain legislation, administrative sanctions may be sufficient or more effective than judicial sanctions;
- an administrative sanction can often be decided and executed without delay, therefore avoiding lengthy and difficult procedures and addressing an immediate risk for the consumer;
- administrative proceedings offer a broader range of sanctions: fines, seizure or destruction of products and suspension/withdrawal of EU agreements; and
- criminal prosecution can be sought for major files but requires solid dossiers since they are competing with other highly pressing criminal investigations such as human trafficking, gun trafficking, murder and terrorism.

The European Commission carries out systematic screenings of all notifications reported in the AAC system and in the RASFF (see also Section 6.2.4). The objective is to identify a potential intentional violation of the EU agri-food chain legislation which may have remained undetected by the Member States and potential violations of the EU agri-food chain which might need adequate coordination and follow up at EU level. These notifications are identified in the RASFF to the attention of the EU countries for further reaction or transferred directly to the EU Food Fraud Network for further treatment and action. Stakeholders are also informed when new or unexpected fraudulent schemes are detected. The Commission services, at the request of one or more EU countries or by its own initiative, can coordinate activities at the EU level. This happens when operators in several EU countries are involved in a possible fraudulent scheme, when operators in non-EU countries might be involved and when the suspicion presents either a health risk or a significant socio-economic risk. Decision criteria for such coordination at the EU level take into account the seriousness of the involved risks, the reliability of the information available and its similarity to previous occurrences. When the suspicion is related to imported products, the Commission works with the concerned non-EU countries and asks for targeted information and investigations. Very concrete results have been achieved in the fight against food fraud thanks to the cooperation between EU countries in the framework of the AAC and between EU countries and non-EU countries, also thanks to consumer awareness and industry vigilance, with the scientific support from the Knowledge Centre for Food Fraud and Quality – Joint Research Centre and the European Food Safety Authority (EFSA) (see the European Commission website available online).

6.7 UNFAIR COMMERCIAL PRACTICES

In general, the 2005/29/EC Directive on unfair commercial practices and its application to food-related consumer protection, which was adopted on 11 May 2005, is considered to provide a good level of consumer protection in a number of sectors and works as a safety net that fills some gaps, which are not regulated by other EU sectors specific rules (Vaqué, 2015). It can be revisited in the light of several Community documents:

- the Commission Guidance Document entitled "Guidance on the implementation/application of Directive 2005/29/EC";
- the Commission's 2013 Communication to the European Parliament, the Council and the European Economic and Social Committee, entitled "Achieving a High Level of Consumer Protection. Building Trust in the Internal Market";
- the Commission's 2013 First Report, entitled "First Report on the Application of Directive 2005/29/EC", amending Council Directive 84/450/EEC, Directives 97/7/EC, 98/27/EC and 2002/65/EC and Regulation (EC) 2006/2004 of the European Parliament and the Council;
- the 2014 European Parliament Resolution on the implementation of the UNFSIR Commercial Practices Directive 2005/29/EC; and
- the 2014 Opinion of the European Economic and Social Committee on Consumer Vulnerability in business practices in a single market.

7

Intentional and Unintentional Substances and Agents Present in Foods

7.1 INTRODUCTION

Maximum levels in different food products of intentionally added substances and agents to food and feed such as additives, flavourings and enzymes and unintentionally added substances and agents such as chemical, biological and physical contaminants are, in general, regulated one by one at the EU level through a procedure that is based on:

- the adoption of a Regulation based on the listing of the authorized individual substances and their maximum levels in different food/feed categories, based on case by case risk assessment; and
- the adoption of a Regulation based on listing biological, chemical and physical contaminants and their maximum tolerable levels in specific food/feed categories by *ad hoc* Regulations, based on case-by-case risk assessment.

7.2 INTENTIONALLY ADDED SUBSTANCES IN FOOD PRODUCTS

The safety evaluation in the EU of flavourings, contact materials, enzymes and processing aids in food and its evolution over time has been reviewed by Silano and Rossi (2014), that of flavourings by Silano (2017), whereas the safety evaluation of food supplements has been reviewed by Silano et al. (2011) and Silano and Fiorani (2016).

7.2.1 Additives

Food additives are substances added intentionally to foodstuffs to perform certain technological functions, for example colouring, sweetening or preserving foods. In the EU, all food additives are identified by an "E number" and are always included in the ingredient lists of foods in which they are used. Product labels must identify both the function of the additive in the finished food (e.g. colour or preservative) and the specific substance used by referring to either the appropriate E number or its name (e.g. E 415 or Xanthan gum). The most common additives to appear on food labels are antioxidants (to prevent deterioration caused by oxidation), colours, emulsifiers, stabilizers, gelling agents and thickeners, preservatives and sweeteners. All additives in the EU can be used only if authorized and listed with conditions of use in the EU positive list based on:

- the safety assessment;
- the technological need; and
- the assurance that the use of the additive will not mislead consumers.

Regulation EC 1333/2008 sets the rules on food additives: definitions, conditions of use, labelling and procedures through:

- Annex I: Technological functions of food additives;
- Annex II: Union list of food additives approved for use in food additives and conditions of use;
- Annex III: Union list of food additives approved for use in food additives, food enzymes and food flavourings, and their conditions of use;
- Annex IV: Traditional foods for which certain EU countries may continue to prohibit the use of certain categories of food additives; and
- Annex V: Additives labelling information for certain food colours.

The EC updates the EU lists of food additives through the regulatory procedure with scrutiny (Decision 1999/468/EC). Producers must inform the Commission of new data which may affect the safety assessment of the food additives already authorized. The list of additives authorized for use in food additives, food enzymes, food flavourings and nutrients can be found in the Annex of Commission Regulation (EU) 1130/2011, which amends Annex III to Regulation (EC) 1333/2008. Food additives must comply with specifications which should include information to adequately identify the food additive, including its origin, and to describe the acceptable criteria of purity. Regulation (EU) 231/2012 laid down specifications for food additives listed in Annexes II and III to Regulation (EC) 1333/2008. Guidance documents on additives are (Paoletti et al., 2012):

- The guidance notes on the classification of food extracts with colouring properties;
- The guidance document describing the food categories; and
- Part E of Annex II to Regulation (EC) 1333/2008 on food additives – the guidance document describing the food categories was updated by the Commission services in February 2016 after consultation with the EU countries' experts on food additives and the relevant stakeholders.

Categories of food additives currently in use in the European Union are listed in Table 7.1.

7.2.2 Flavourings

Regulation (EC) 1334/2008 on flavourings and certain food ingredients with flavouring properties for use in/on foods was adopted on 16 December 2008 and entered into force on 20 January 2009. This Regulation lays down general requirements for safe use of flavourings and provides definitions for different types of flavourings. The Regulation sets out substances for which an evaluation and approval is required. The Union list of flavouring substances, approved for use in and on foods, was adopted on 1 October 2012 and was introduced in Annex I of Regulation (EC) 1334/2008 (Paoletti et al., 2011b). Since then, the Union list of flavourings has been amended many times (Table 7.2).

The available Regulations prohibit the addition of certain natural undesirable substances as such to food and lay down maximum levels for certain substances, which are naturally present in flavourings and food ingredients with flavouring properties, but which may raise concern for human health. The overall rules on the labelling of food are also relevant for flavourings. However, specific provisions regarding the labelling of flavourings are also in Chapter IV of Regulation (EC) 1334/2008. Moreover, Regulation (EC) 1334/2008 also sets out the rules for flavourings from business to business

Table 7.1 Categories of Food Additives Currently in Use in the European Union

Colours: substances giving colour to foods.

Preservatives: substances slowing down or preventing the growth of microorganisms in food.

Antioxidants: substances preventing oxidation that causes rancid fats and brown fruits.

Acids and acidity regulators: substances modifying the acidity of foods.

Emulsifiers: substances facilitating the mixing of water and oils in emulsions.

Anti-caking agents: substances preventing the agglomeration of foods.

Anti-foaming agents: substances preventing or reducing foaming.

Firming agents: substances that help keeping firm or crispy structures of foods.
Flavour enhancers: substances enhancing the existing flavours/odours of food.

Flavourings: substances imparting or modifying the food odour and/or taste.

Foaming agents: substances with a surfactant activity which when present in small amounts, facilitate the formation of a foam or enhance its stability.

Thickeners and gelling agents: substances improving the textures of foods by increasing the viscosity/thickness or by forming a gel.

Bulking agents: substances increasing the volume or weight of foods without significantly changing its caloric content.

Stabilizers: substances maintaining the physical and chemical state of a food.

Sweeteners: low-calorie substances giving a sweet taste to foods.

Glazing agents: substances imparting a shiny appearance of the outer surface of food or providing a protective coating.

Humectants: substances that facilitate keeping the original dry matter to moisture ratio in foods.

Modified starches: starch-like carbohydrates obtained by the treatment of corn or wheat starch with heat, alkali, acids or enzymes that, in the human intestine, may act as soluble or insoluble dietary fibre, more or less fermentable and only partially digestible and provide less calories than regular starch.

Packaging gases: substances used to replace the natural ambient air in the package, e.g. to help preserving the quality of foods.

Propellants: substances helping to expel the food from its container.

Leavening/raising agents: substances impacting baked goods by their ability to form CO_2 in the baking process.

Flour treatment agents: substances used to increase the speed of dough rising and to improve the strength and workability of the dough.

Table 7.2 Commission Regulations on Flavourings Adopted between 2013 and 2019

Commission Regulation (EU) 545/2013 of 14 June 2013
Commission Regulation (EU) 985/2013 of 14 October 2013
Commission Regulation (EU) 246/2014 of 13 March 2014
Commission Regulation (EU) 1098/2014 of 17 October 2014
Commission Regulation (EU) 2015/648 of 24 April 2015
Commission Regulation (EU) 2015/1102 of 8 July 2015
Commission Regulation (EU) 2015/1760 of 1 October 2015
Commission Regulation (EU) 2016/54 of 19 January 2016
Commission Regulation (EU) 2016/55 of 19 January 2016
Commission Regulation (EU) 2016/178 of 10 February 2016
Commission Regulation (EU) 2016/637 of 22 April 2016
Commission Regulation (EU) 2016/692 of 4 May 2016
Commission Regulation (EU) 2016/1244 of 28 July 2016
Commission Regulation (EU) 2017/378 of 3 March 2017
Commission Regulation (EU) 2017/1250 of 11 July 2017
Commission Regulation (EU) 2018/678 of 3 May 2018
Commission Regulation (EU) 2018/1246 of 18 September 2018
Commission Regulation (EU) 2018/1482 of 4 October 2018
Commission Regulation (EU) 2018/1649 of 5 November 2018
Commission Regulation (EU) 2019/36 of 10 January 2019
Commission Regulation (EU) 2019/799 of 17 May 2019

and for sale to the final consumers. It also describes the specific requirements for the use of the term "natural". This Regulation also harmonizes and clarifies the rules for the use of these substances in the EU.

Smoke flavourings are a specific category of flavourings which are governed by a specific set of Regulations:

- Regulation (EC) 2065/2003 of the European Parliament and of the Council of 10 November 2003 on smoke flavourings used or intended for use in or on foods. The Regulation establishes a Community procedure for the safety assessment and the authorization of smoke flavourings intended for use in or on foods in order to ensure a high level of protection of human health and protection of consumers' interests, as well as to ensure fair trade practice;

- Regulation (EC) 627/2006 of 21 April 2006 implementing Regulation (EC) 2065/2003 of the European Parliament and of the Council as regards quality criteria for validated analytical methods for sampling, identification and characterization of primary smoke products; and
- Regulation (EU) 1321/2013 establishing the Union list of authorized smoke flavouring primary products for use as such in or on foods and/or for the production of derived smoke flavourings was published in *Official Journal* on 12 December 2013. This Regulation lists the authorized smoke flavouring primary products for use in or on foods and their conditions of use. This list was established on the basis of the applications submitted under Article 20 of Regulation (EC) 2065/2003 and after evaluation by the European Food Safety Authority (EFSA).

The food additives that may be used in flavourings and their conditions of use are specified in the Regulation on additives:

- Older legislations on flavourings are still partially valid. This is the case for European Parliament and Council Regulation (EC) 2232/96 of 28 October 1996 laying down a Community procedure for flavouring substances used or intended for use in or on foodstuffs. This Regulation is expected to remain in force until a number of substances for which the EU risk assessment has not been completed at the time of entry into force of Regulation (EU) 872/2012 have been fully evaluated and added to the Union list. At that time, the authorization procedure for flavourings under Regulation (EC) 1331/2008 will become fully applicable.
- Regulation (EC) 1565/2000 of 18 July 2000 laying down the measures necessary for the adoption of an evaluation programme in the application of Regulation (EC) 2232/96 of the European Parliament and of the Council. This Regulation is expected to remain in force until a number of substances for which the EU risk assessment has not been completed at the time of entry into force of Regulation (EU) 872/2012 have been fully evaluated and added to the Union list. At that time, the authorization procedure for flavourings under Regulation (EC) 1331/2008 becomes fully applicable.

7.2.3 Extraction Solvents

Extraction solvents are solvents used in an extraction procedure during the processing of raw materials, foodstuffs, or components or ingredients of these products and which are removed but which may result in the unintentional, but technically unavoidable, presence of residues or derivatives in the foodstuff or food ingredient.

Solvent means any substance for dissolving a foodstuff or any component thereof, including any contaminant present in or on that foodstuff. The EU rules on extraction solvents for use in foodstuffs should take into account primarily human health requirements and also within the limits required for the protection of health, economic and technical needs. Directive 2009/32/EC on the approximation of the laws of EU countries on extraction solvents used in the production of foodstuffs and food ingredients applies to extraction solvents used or intended for use in the production of foodstuffs or food ingredients either in the EU or imported into the EU. This Directive does not apply to:

- extraction solvents or foodstuffs intended for export outside the EU; and
- extraction solvents used in the production of food additives, vitamins and other nutritional additives, unless such food additives, vitamins or nutritional additives are listed in Annex I of the Directive.

The EU Member States shall not authorize the use of other substances and materials as extraction solvents, nor they shall extend the conditions of use or permitted residues of the extraction solvents listed in Annex I beyond those specified therein. For amending Directive 2009/32/EC, it is recommended to follow the requirements set in Regulation (EU) 234/2011. It is essential that the technological need be well explained, including the reason for setting a maximum residual level at a certain level. It should also be clearly demonstrated that the (new) extraction solvent (use) would not imply any safety concern.

7.2.4 Food Enzymes

Regulation (EC) 1332/2008 on food enzymes harmonized the rules on food enzymes for the first time in the EU and fixed a deadline of two years for the submission of applications for authorization. This Regulation deals only with food enzymes added to food for a technological purpose in the manufacture, processing, preparation, treatment, packaging, transport or storage of food, including enzymes used as processing aids.

Enzymes not covered by this Regulation include:

- enzymes intended for human consumption, e.g. for nutritional or digestive purposes; and
- food enzymes used in the production of food additives under Regulation (EC) 1333/2008 and in the production of processing aids.

Microbial cultures traditionally used in the production of food (e.g. cheese and wine), which may incidentally produce enzymes but are not specifically used to produce them, are not considered food enzymes. Meanwhile, the Commission reached the conclusion that the initial deadline for submitting applications was insufficient in order to allow stakeholders and, in particular, small- and medium-size enterprises to produce all necessary data within that period. Therefore, the 24-month period was extended to 42 months by Commission Regulation (EU) 1056/2012. The safety of food enzymes is being evaluated by EFSA and will be approved by "comitology" procedure (establishing the EU list). Until the EU list of food enzymes is drawn up, national rules on the marketing and use of food enzymes and food produced with food enzymes will continue to apply in EU countries. Other pieces of EU legislation relevant to food enzymes are the following:

- Regulation (EC) 1331/2008 – establishes the common authorization procedure for food additives, food enzymes and food flavourings; and
- Regulation (EU) 234/2011 – implements the common authorization procedure. This Regulation has been amended by Commission Implementing Regulation (EU) 562/2012, which lays down derogation from submitting toxicological data in some specific cases and the possibility of grouping food enzymes under one application under certain conditions.

A specific "guidance document" has been produced by the Commission services after consultation with the EU countries' experts on food enzymes and the relevant stakeholders with the aim of providing informal guidance for food business operators (FBOs) and competent authorities on criteria for determining the status of a food enzyme either as an ingredient or as a processing aid in a given context of use, and hence whether it needs to be listed in the ingredient list of foods intended for the final consumer. Such criteria will help applicants to prepare an appropriate application for the authorization of food enzymes. The guidance also includes a decision tree to facilitate this categorization. This Guidance document was endorsed by a majority of the representatives of the EU countries at the meeting of the Standing Committee on the Food Chain and Animal Health on 20 February 2014. A statement on dietary exposure to food enzymes was adopted in 2016 after public consultation and presentation to stakeholders during an info-session and an *ad hoc* meeting with the Association of Manufacturers and Formulators of Enzymes Products (AMFEP). The statement is a food process-based methodology to estimate dietary exposure to food enzymes that require additional data to be submitted by the applicants for different food processes.

EFSA launched open calls to collect input data to allow the implementation of the methodology and requested technical specifications to the enzyme manufacturers association (AMFEP). The development of web-based food enzyme intake models (FEIMs) was started as a self-task activity. FEIM calculations help estimate dietary exposure to food enzymes used in individual food processes, and two FEIMs have been published (covering baking and brewing), while two are under production (savoury snack processing and cereal-base processes) and many more are planned.

So far, EFSA has received many dossiers on food enzymes for authorization and a multi-annual work programme for evaluations of enzymes based on food processes has been discussed and agreed with the European Commission. Forty-two opinions have been already adopted by the CEP Panel, and it is obvious that food enzymes will be a priority working area for the EFSA CEP Panel for several years.

7.2.5 Processing Aids Including BPA

The term "processing aid" is defined in Regulation (EC) 1333/2008 on food additives as any substance that: (i) is not consumed as a food by itself; (ii) is intentionally used in the processing of raw materials, foods or their ingredients, to fulfil a certain technological purpose during treatment or processing; and (iii) may result in the unintentional but technically unavoidable presence in the final product of residues of the substance or its derivatives, provided they do not present any health risk and do not have any technological effect on the final product. The term "processing aids" is defined also by the *Codex Alimentarius* (in the Procedural Manual) as any substance or material, not including apparatus or utensils, and not consumed as a food ingredient by itself, intentionally used in the processing of raw materials, foods or its ingredients to fulfil a certain technological purpose during treatment or processing and which may result in the non-intentional but unavoidable presence of residues or derivatives in the final product. There are Codex Guidelines on substances used as processing aids. At European level no specific rules for enforcement and controls of processing aids are available, and the general principles of the food safety legislation apply. FBOs at all stages of production, processing and distribution within the business under their control shall ensure that foods satisfy the requirements of food laws which are relevant to their activities and shall verify that such requirements (e.g. traceability and HACCP) are met. The Member States shall enforce food laws and monitor and verify that the relevant requirements of food laws are fulfilled by FBOs at all stages of production, processing and distribution.

Bisphenol A (BPA) is a chemical substance mainly used in the manufacturing of plastics and resins. The safety of BPA has been evaluated by EFSA in 2006, 2008 and 2010. In 2015 a new evaluation of BPA safety was carried out, and the previously adopted TDI was temporarily reduced to a TDI = 4 µg/kg b.w. and day due to existing uncertainties. In 2016, following the request from a Member State to assess two new studies, a statement was adopted by the CEF Panel on BPA developmental immunotoxicity, but no further revision of t-TDI was considered as necessary.

The need was recognized for developing a protocol for reviewing all scientific evidence published on BPA since 2013 and especially associated with low BPA doses on mammary glands, reproductive, neurological, immune and metabolic systems.

A call for data took place from March till October 2018 to gather human and animal hazard studies/data that have been published since 2013 and that are relevant to BPA hazards. A working group on BPA, started in 2018, is currently active on re-evaluating BPA toxicity.

7.2.6 Plant Protection Products and Biocides

Substances used to suppress, eradicate and prevent actions of organisms that are considered harmful are grouped under the term "pesticides". This term includes both plant protection products (used on plants in agriculture, horticulture, parks and gardens) and biocidal products (used in other applications, for example, as a disinfectant or to protect materials). Pesticides can be useful in a number of circumstances, for example in overcoming diseases and increasing agricultural yields. However, their environmental impacts and the risks that they may pose to human health may also induce reasons for concern.

7.2.6.1 Plant Protection Products

Plant protection products (PPPs) are products in the form in which they are supplied to the user, consisting of, or containing, active substances, safeners or synergists, and intended for one of the following uses:

a. protecting plants or plant products against all harmful organisms or preventing the action of such organisms, unless the main purpose of these products is to promote hygiene rather than to protect plants or plant products (e.g. fungicides);

b. influencing the life processes of plants, such as substances influencing their growth, other than as a nutrient (e.g. plant growth regulators, rooting hormones);

c. preserving plant products, in so far as such substances or products are not subject to special Community provisions on preservatives (e.g. extending the life of cut flowers); and

d. destroying undesired plants or parts of plants or checking or preventing undesired growth of plants, except algae, unless the products are applied on the soil or water to protect plants (e.g. herbicides/weedkillers to kill actively growing weeds).

PPPs are subject to a dual approval process: active substances are approved at the EU level, and specific products are subsequently authorized predominantly at the Member State level. Furthermore, standardized maximum levels are set for the residues of PPPs in food, and a framework for action is focused on sustainable pesticide use. A number of EU Regulations have been adopted on PPPs (Table 7.3).

Regulation (EC) 1107/2009 deals with the rules and procedures for the placing of PPPs on the market in the European Union. This Regulation lays down the controls of the use and placing on the market of PPPs that are performed by EU countries. The Plant Protection Products Application Management System (PPPAMS) was developed by the European Commission to enable industry users to create applications for PPPs and submit these to EU countries for evaluation. EU countries then manage these applications within the system, concluding with the authorization of the PPP or refusal of the application. The system is designed to support EU countries in fulfilling their legal obligations under Regulation (EU) 1107/2009, notably Articles 57(1) and (2).

The objectives of the PPPAMS are:

• harmonization of the formal requirements for the application of PPPs among EU countries;

Table 7.3   EU Regulations Adopted on Plant Protection Products

Regulation CE 1107/2009 – placing on the market of PPPs

Regulation UE 540/2011 – approved substances (updated on 03/09/2014)

Regulation UE 546/2011 – uniform principles for the evaluation and the
  authorization of PPPs

Regulation UE 547/2011 – labelling requirements for PPPs

Regulation UE 283/2013 – definition of requirements concerning data for active
  substances

Regulation UE 284/2013 – definition of requirements concerning data for PPPs

Relation on minor use (Article 51)

Questions and answers (Updated on 07/12/2015)

Directive 91/414/CEE – evaluation, authorization and approval of active
  substances at the UE level and national authorization of PPPs

Regulation CE 33/2008 – fulfilment norms of Directive 91/414/CEE – regular
  and accelerated procedure for the evaluation of active substances

Regulation CE 1095/2007 – fulfilment norms of Directive 91/414 – CEE stages 1,
  2, 3 and 4 of the work programme to examine active substances

Regulation CEE 3600/92 – fulfilment of the first stage

Regulation CE 451/2000 – fulfilment of the second stage

Regulation CE 1490/2002 – fulfilment of the third stage

Regulation CE 2229/2004 – fulfilment of the fourth stage

Regulation UE 188/2011 – procedure for evaluating new active substances

Source: European Commission Website.

- streamlining the mutual recognition of authorizations for PPPs among EU countries to speed up time to market;
- improving the management of the evaluation process for the authorization of PPPs; and
- delivering correct and timely information on authorized or withdrawn plant protection products to stakeholders

The system allows applicants to create and manage applications and submit them to EU countries for evaluation. Authorizations issued by EU countries are stored in the system and made available through a database.

EU legislation harmonizes the adoption of pesticide maximum residue levels (MRLs) and sets a common EU assessment scheme for all agricultural products for food or animal feed:

- MRLs apply to 315 categories of fresh products and to the same products after processing, adjusted to take account of dilution or concentration during the process;
- legislation covers pesticides currently or formerly used in agriculture in or outside the EU (around 1,100);
- a general default MRL of 0.01 mg/kg applies where a pesticide is not specifically mentioned;
- the safety of all consumer groups is covered, e.g. babies, children and vegetarians;
- EFSA assesses the safety for consumers based on the toxicity of the pesticide, the maximum levels expected in food and the different diets of Europeans; and
- clear roles of the Member States, EFSA and the Commission in the setting of MRLs are well defined.

Applicants, e.g. producers of plant protection products, farmers, importers, EU or non-EU countries, must submit the following for the setting of MRLs for pesticides:

- the use of a pesticide on the crop, e.g. quantity, frequency, the growth stage of the plant (good agricultural practice – GAP);
- experimental data on the expected residues when the pesticide is applied according to GAP; and
- toxicological reference values for the pesticide. Chronic toxicity is measured with the acceptable daily intake (ADI) and acute toxicity – with the acute reference dose (ARfD).

Based on the available information, the intake of residues through all food which may be treated with that pesticide is compared with the:

- ADI; and
- ARfD for long- and short-term intake and all European consumer groups.

If the requested MRL is not safe, it is set at the lowest limit of analytical determination (LOD). That is also the MRL for crops on which the pesticide has not been used or when its use has not left detectable residues. The default lowest limit (LOD) in EU law is 0.01 mg/kg.

7.2.6.2 Biocides

In the case of biocides, Regulation (EU) 528/2012 sets out rules for:

- approving active substances in biocidal products;
- authorizing the supply and use of biocidal products; and
- the supply of articles treated with biocidal products.

In particular, it states that a biocidal product cannot be placed on the market or used unless it contains approved active substances and has been preliminarily authorized. This Regulation also includes provisions to reduce animal testing by making data sharing on vertebrate studies compulsory and encouraging a more flexible and intelligent approach to testing.

The Regulation outlines the general principles and products which fall under the scope of the biocidal products Regulation, including mixtures, articles and materials treated with biocidal products, furniture and textiles, as well as a provision on dual use covering biocidal products that have a dual function.

The applicable definitions are also outlined, making a distinction between those substances, mixtures and articles that should be regarded as biocidal products and those which should not, differentiating notably between them.

Moreover, although the Regulation does not ban animal testing completely, it attempts to minimize it as much as possible. Article 62 introduces an obligation to share data on vertebrate animal tests in exchange for fair compensation and a prohibition to duplicate such tests, which is aimed at saving costs, as well as animal lives.

It also encourages data sharing for non-animal tests, with a similar view to reducing the overall costs of the legislation to the industry and avoiding the duplication of efforts. Where necessary, the European Chemicals Agency (ECHA) is involved in the data sharing process.

To obtain the authorization needed to supply and use these products, companies must demonstrate that the product is effective and does not present unacceptable risks to humans, animals or the environment. Interested companies can choose among different approaches for getting the authorization of biocides, depending on the nature of the product and the number of countries in which the product is intended to be marketed. If the product is intended for a single Member State market, the authorization of the specific country is sufficient. In case a company intends to market the product already authorized in a specific Member State in different states, it can apply for a mutual recognition in all the Member States of interest. For the renewal of the national authorization and mutual recognition, the same procedure applies. For interested companies it is also possible to apply for an authorization at the level of the European Union in a unique procedure. In these cases, it is the European Commission that grants such Union authorizations.

For products which satisfy specific requirements detailed in the Regulation (e.g. being devoid of substances of concern), there is also the possibility of applying for a simplified authorization. The ECHA is responsible for providing technical and scientific support in implementing Regulation (EU) 528/2012. Through its Biocidal Product Committee, it provides opinions to the European Commission, among others, on:

- approving active substances; and
- authorizing biocidal products at the EU level.

On the basis of ECHA's opinions, the Commission then decides whether to approve active substances and grant Union authorizations to biocidal products. ECHA also provides secretariat support to the Coordination Group, which plays an important role in the process of mutual recognition of product authorizations at the national level between EU countries and is responsible for maintaining the Register for Biocidal Products, an IT system used to:

- submit applications under the above-mentioned Regulation;
- exchange information while applications are being assessed; and
- disseminate information once active substances have been approved and products have been authorized.

ECHA also manages the list of suppliers of active substances established under Article 95 and related applications. ECHA further assesses technical equivalence and data sharing inquiries.

7.3 UNINTENTIONALLY PRESENT SUBSTANCES IN FOOD PRODUCTS

7.3.1 Chemical Contaminants

Contaminants are substances that have not been intentionally added to food. These substances may be present in foods as a result of the various stages of its production, packaging, transport or holding. Specific contaminants may also result from environmental contamination. Since contamination generally has a negative impact on the quality of food and may imply a risk to human health, the EU has taken measures to minimize contaminants in foodstuffs (Silano,1997; Silano and Silano, 2015).

The basic principles of EU legislation on contaminants in food are laid down in Regulation (EEC) 315/93 that does not apply to contaminants such as pesticide residues and veterinary drug residues which are the subjects of more specific Community rules (see the previous section and Audino et al., 1995):

- food containing a contaminant to an amount unacceptable from the public health viewpoint and in particular at a toxicological level shall not be placed on the market;
- contaminant levels shall be kept as low as it can reasonably be achieved following recommended good working practices;
- maximum levels must be set for certain contaminants in order to protect public health;
- sampling and analysis methods shall be developed and implemented;
- consultation of EFSA is expected before maximum levels are set;
- regulatory decisions (vote) shall be taken by the Standing Committee on the Food Chain and Animal Health.

The European Commission published (2008) a factsheet on food contaminants: "Managing Food Contaminants: How the EU Ensures That Our Food Is Safe". Maximum levels in food have been set with Regulation (EC) 1881/2006 and subsequent amendments for the following contaminants:

- nitrate;
- mycotoxins (aflatoxins, ochratoxin A, patulin, deoxynivalenol, zearalenone, fumonisins and citrinine);
- metals (lead, cadmium, mercury, inorganic tin and arsenic);
- chloropropanols (e.g. 3-MCPD);
- dioxins (Silano and Comba, 1989);
- dioxin-like PCBs;
- non-dioxin-like PCBs;
- polycyclic aromatic hydrocarbons (PAH) (benzo(a)pyrene) and a sum of four PAHs);
- melamine; and
- erucic acid.

When setting maximum limits of food contaminants, the ALARA principle for food contaminants should be carefully considered. Therefore, the maximum level of contaminants should be as low as reasonably achievable and technologically practicable and based on the ability to analyze the contaminant in the food of interest. The balance between toxicological and other factors, including social, technical and economic factors, has also to be taken into account. In essence, the maximum levels of food contaminants have to be established following scientific risk assessment (as carried out by EFSA). The monitoring of food contaminants follows a similar process for all contaminants and focusses by means of statistical/ scientific principles on:

- sampling at the earliest possible point in the food chain; and
- targeting at import/wholesale levels.

With the aim to capture general background levels, several EU Regulations have been adopted on the monitoring of specific contaminants:

- Regulation (EC) 1883/2006 on dioxins;
- Regulation (EC) 1882/2006 on nitrates;
- Regulation (EC) 333/2007 on metals, inorganic tin and PAHs; and
- Regulation (EC) 401/2006 on aflatoxins, ochratoxin A, patulin and fusarium toxin.

For some contaminants, specific regulations establishing mitigation measures have been adopted at the European level. This is the case of acrylamide and the Commission Regulation (EU) 2017/2158 of 20 November 2017 establishing mitigation measures and benchmark levels for the reduction of the presence of acrylamide in food. As the acrylamide levels in some foodstuffs appear to be significantly higher than the levels in comparable products of the same product category, a Commission Recommendation 2013/647/EU invited Member States' competent authorities to carry out investigations in the production and processing methods used by FBOs if the acrylamide level found in a specific foodstuff exceeded the indicative values set out in the Annex to that Recommendation. Moreover, in 2015 the EFSA Scientific Panel CONTAM adopted an opinion on acrylamide in food, confirming previous evaluations that acrylamide in food potentially increases the risk of developing cancer for consumers in all age groups. Since acrylamide is present in a wide range of everyday foods, this concern applies to all consumers, but children are the most exposed age group on a body weight basis. The current levels of dietary exposure to acrylamide across age groups indicate a concern with respect to its carcinogenic effects.

7.3.2 Biological Contaminants

Living microorganisms can be hazardous to human health if swallowed or otherwise absorbed into the body. Biological hazards are characterized by the contamination of food by microorganisms. Found in the air, food, water, animals and the human body, foodborne illness can occur if harmful microorganisms make their way into the food we eat. There are several types of microorganisms which can negatively impact health: bacteria, viruses and parasites. Biological contamination is one of the common causes of food poisoning, as well as spoilage. During biological contamination, the harmful bacteria spread on foods. Even a single bacterium can multiply very quickly under ideal growth conditions. Microbiological criteria are used to assess the acceptability of food. When a particular food is tested for a particular microorganism (toxin or metabolite) the results can indicate whether:

- the food is safe to eat or not;
- the food is of acceptable quality; or
- the hygiene standards in the food establishment are satisfactory or unsatisfactory.

Legal microbiological criteria have been set for some, but not all, combinations of food and microorganisms. If a relevant legal microbiological criterion does not exist, microbiological guideline criteria can be used to assess the acceptability of food. The main objectives of microbiological criteria and their harmonization include:

- ensuring a high level of human health protection;
- reducing human cases of foodborne diseases;
- being founded on scientific risk assessments;
- being feasible in practice;
- verifying the HACCP measures; and
- standardizing rules within the EU and for the importation.

Components of microbiological criteria include:

- microorganisms of concern;
- analytical method;
- sampling plan;
- the number of sample units;
- size of the analytical unit;
- microbiological limits;

- foodstuff;
- the point of the food chain where the limit applies; and
- actions to be taken when unsatisfactory results are obtained.

The responsibilities of the FBOs include:

- demonstration of the compliance with microbiological criteria and food safety criteria throughout the shelf-life of foods;
- durability, challenge studies;
- establishment of a sampling and testing scheme based on risk (HACCP);
- response in case of non-compliance; and
- follow-up and assessment of trends.

Regulation (EC) 2073/2005 harmonized the microbiological food safety and process hygiene criteria for foodstuffs in the EU. The process hygiene criteria (Table 7.4) are:

- used to assess the correct functioning of production processes. They are applicable to foodstuffs either during or at the end of the manufacturing process; and
- established for microorganisms (usually indicator microorganisms) in various food commodities, e.g. E. coli in minced meat, Enterobacteriaceae in egg products.

The microbiological food safety criteria (Table 7.5) define the acceptability of the food products placed on the market.

In case of unsatisfactory results, the possible actions include:

- food withdrawal or recall; and
- food further processing (in case food is not yet at the retail level) and other corrective actions based on the HACCP programme.

Table 7.4 Examples of Process Hygiene Criteria

Salmonella: carcases

Aerobic colony count and Enterobacteriaceae: carcases

Escherichia coli: minced meat, meat preparations, butter and cream made from raw milk, precut fruit and vegetables

Coagulase-positive *Staphylococci*: certain dairy products

Enterobacteriaceae: Dried infant formulae

Table 7.5   Examples of Food Safety Criteria

Listeria monocytogenes: all ready-to-eat foods
Salmonella: certain ready-to-eat foods, minced meat, meat preparations and meat products
Staphylococcal enterotoxins: certain dairy products
E. sakazakii: dried infant formulae
E. coli: live bivalve mollusks
Histamine: fishery products from certain fish species

Source: European Commission Website.

The food safety criteria for *Salmonella* are as follows: minced meat, meat preparations and meat products intended to be eaten raw and cooked. Raw: absence in 25 g, n = 5, c = 0. Cooked: absence in 10 g, n = 5, c = 0. Commission Regulation (EC) 2073/2005 on microbiological criteria was amended by Regulation (EC) 1441/2007 of 5 December 2007 with criteria on Enterobacteriaceae and *Salmonella* in dried follow-on formulae and *Bacillus cereus* in dried infant formulae and the reference method for the staphylococcal enterotoxin detection. Also relevant in this context is the "guidance document" on official controls, under Regulation (EU) 882/2004, concerning microbiological sampling and testing of foodstuff. From the practical point of view, it can be concluded that the EU Member States can have different (national) microbiological food safety criteria, only if they are stricter than those adopted at the European level.

7.3.3 Physical Contaminants

Physical contamination is any visible foreign object found in food. Human hair is the most commonly occurring physical contaminant of food. Other common examples of physical contaminants include metal, glass or animal parts not originally present in the food source. Some of these hairs end up in food when hygiene standards in food processing plants are not strictly enforced. Animal parts also typically become physical contaminants in food when proper cleanliness is not maintained. In home kitchens, broken glass and broken knife tips are typically the causes of physical food contamination. Since it is impossible to tell if physical food contamination has also caused chemical contamination or bacterial contamination, food in which foreign objects are found should not be consumed. A large number of food products are recalled each year due to foreign contaminants found by consumers. These recalls not only cost companies a lot of money, but they can damage company reputations and put consumer health at risk. Some real-life examples from the U.S. Food and Drug Administration (FDA) 2016 product recall list shows that almost every packaged food is susceptible to foreign contaminant problems. The food industry takes many precautions to ensure that any food that reaches consumers is free of physical contaminants like metal, glass and stone, which can enter a product or package anytime during farming or processing. There are X-ray inspection systems that use X-rays and sensors and food metal detectors that use metal coils and high frequencies to find contaminants before they reach consumers. New food metal detectors are equipped with a multiscan technology that enables operators to pick a set of up to five frequencies from 50 kHz to 1,000 kHz. The technology then scans through each frequency at a very rapid rate, effectively acting like five metal detectors in one.

7.3.4 Radioactive Contaminants

The environment and, consequently, food contain measurable amounts of radionuclides arising primarily from the distribution of nuclear bomb debris over the entire earth's surface. This radioactivity is in addition to natural radioactivity, which exists in the environment. Additional and more dangerous sources of radioactivity are represented by accidents at nuclear plants, which may cause the release of a radioactive fallout that may contaminate agricultural foods directly or through the soil (Campos Venuti et al., 1985). Very relevant in this context is the Council Regulation (Euratom) 2016/52 of 15 January 2016 laying down maximum permitted levels of radioactive contamination of food and feed following a nuclear accident or any other case of a radiological emergency, and repealing Regulation (Euratom) 3954/87 and Commission Regulations (Euratom) 944/89 and (Euratom) 770/90.

Moreover, in the event of a radiological emergency, if one Member State takes measures to protect the general public, a notification is due to the European Commission of the action taken. The information must include the nature and time of the event, its exact location and the nature of the facility or activity involved, the cause, the foreseeable development and the protective measures taken or planned. The other Member States that may be affected are required to advise the Commission of the levels of radioactivity measured by their monitoring facilities in foodstuffs, feedstuffs, drinking water and the environment.

7.3.5 Food Contact Materials Including
Recycled Plastic Materials and Articles

Food comes into contact with many materials and articles during its production, processing, storage, preparation and serving, before its consumption. Such materials and articles are called food contact materials (FCMs). FCMs are either intended to be brought into contact with food, are already in contact with food or can reasonably be brought into contact with food or transfer their constituents to the food under normal or foreseeable use. Examples include:

- containers for transporting food;
- machinery for processing food;
- packaging materials; and
- kitchenware and tableware.

The term does not cover fixed public or private water supply equipment. FCMs should be sufficiently inert so that their constituents neither adversely affect consumer's health nor influence the quality of the food. To ensure the safety of FCMs, and to facilitate the free movement of goods, EU law provides for binding rules that FBOs must comply with. The EU rules on FCMs can be of general scope, i.e. to apply to all FCMs or to specific materials only. EU law may be complemented with the Member States' national legislation if specific EU rules do not exist. The safety of FCMs is evaluated by EFSA. The safety of FCMs is tested by the business operators interested in placing them on the market and by the competent authorities of the Member States during official controls. Scientific knowledge and technical competence in testing methods are being maintained by the European Reference Laboratory for Food Contact Materials (EURL-FCM). Its website provides guidelines and other resources concerning the testing of food contact materials. Regulation (EC) 1935/2004 sets out the general principles of safety and inertness for all FCMs. The principles set out in Regulation (EC) 1935/2004 require that packaging materials do not:

- release their constituents into food at levels harmful to human health; and
- change food composition, taste and odour in an unacceptable way.

Moreover, the framework provides:

- special rules for active and intelligent materials (they are by their design not inert);

- powers to enact additional EU measures for specific materials (e.g. for plastics);
- the procedure to perform safety assessments of substances used to manufacture FCMs involving the European Food Safety Authority; and
- rules on labelling including an indication for use (e.g. as a coffee machine, a wine bottle, or a soup spoon) or by reproducing the appropriate symbol (for more information, please refer to the document on "Symbols for Labelling Food Contact Materials" for compliance documentation). Moreover, Regulation (EC) 2023/2006 ensures that the manufacturing process is well controlled so that the specifications for FCMs remain in conformity with the legislation through specific requirements such as:
  - premises fit for purpose and staff awareness of critical production stages;
  - documented quality assurance and quality control systems maintained at the premises; and
  - selection of suitable starting materials for the manufacturing process with a view to the safety and inertness of the final articles.

Good manufacturing rules apply to all stages in the manufacturing chain of food contact materials, although the production of starting materials is covered by other legislation. In addition to the general legislation, certain FCMs – ceramic materials, regenerated cellulose film, plastics (including recycled plastic), as well as active and intelligent materials – are covered by specific EU Regulations. There are also specific rules for some starting substances used to produce FCMs. The Scientific Committee for Foods (SCF) guideline for the evaluation of substances used in plastic FCMs was first published in 1990 and updated in 2001. The SCF guideline was endorsed in 2003 by the EFSA AFC Panel and complemented in 2008 by the CEF Panel with an EFSA explanatory note for guidance, with detailed technical information on a core set of toxicity data The "Explanatory note" and "Administrative EFSA guidance" have been updated in 2017. Two regulatory approaches regulate the use of recycled plastics in FCMs. Plastic depolymerized into monomers or oligomers have to meet the same requirements as virgin materials. When plastic is mechanically recycled and transformed into pellets, Regulation EC 282/2008 foresees an individual authorization for the recycling process to be carried out by the EFSA. The EFSA guidance for evaluating recycled plastics was adopted in 2008 and the EFSA criteria for evaluating recycled PET in 2011.

The general task of evaluating substances intended for use in FCMs as well as carrying out additional risk assessments in relation to FCMs are carried out by EFSA CEP Panel. This work forms part of the authorization procedure for substances which require evaluation by EFSA before their use in the EU can be authorized. The panel's work is based on reviewing scientific information and the data usually submitted by applicants. In 2009, EFSA has also adopted a guidance for submission and safety evaluation of applications related to active (able to extend shelf-life or maintain/improve conditions of packaged foods) and intelligent (able to monitor conditions of packaged foods or surrounding environment) materials (AIM).

8

Particular Food Products

8.1 GENETICALLY MODIFIED FOODS

In the EU, a genetically modified food (GMO) is any organism created using genetic engineering techniques. GMOs are produced by recombining the DNA of two or more different organisms with the aim of developing a new organism with more desirable properties. This helps to make genetically modified crops more resistant to certain herbicides, pests, diseases, extreme weather conditions or taste better, last longer on the shelf or to improve their nutritional content.

Initially, genetic engineering is focused on increasing crop yields and making plants easier and cheaper to produce. More recently, genetically altered crops with potential benefits to the consumer have been developed, having the following desired traits:

- improved taste and/or appearance;
- enhanced nutritional value and health;
- improved adaptability to environmental conditions; and
- health benefits.

Genetic modification is also being used to develop crops with other benefits to consumers (e.g., potatoes with higher starch content) as well as benefits to the environment (e.g., plants which absorb certain toxic substances from the soil or water). However, several concerns have also been raised regarding genetically modified foods, including:

- introducing new allergenic proteins in genetically modified foods;

- contributing to the development of antibiotic-resistant strains of bacteria;
- uncontrollable cross-breeding with traditional or wild plants that could allow weeds to develop resistance to pesticides;
- spreading of genetically modified plants beyond controlled areas to become "super weeds";
- pests developing resistance to the toxins produced by genetically modified plants;
- toxins produced by genetically modified plants affecting or killing non-target organisms; and
- potentially mingling of genetically modified crops not approved for human consumption with their conventional counterparts and using them in food production.

In relation to GM foods, the EU differentiates between approval for cultivation within the EU and approval for import and processing. While only a few GMOs have been approved for cultivation in the EU, a number of GMOs have been approved for import and processing. The cultivation of GMOs has triggered a debate about the market for GMOs in Europe. One of the key issues was whether GM products should be labelled. According to the EU, mandatory labelling and traceability are needed to allow for informed choice, avoid potential false advertising and facilitate the withdrawal of products if adverse effects on health or the environment are discovered. In the EU all foods (including processed foods) or feed which contains more than 0.9% of approved GMOs must be labelled. A legal framework has been established in the EU to ensure that the development of modern biotechnology, and more specifically of GMOs, takes place in safe conditions. The legal framework aims to:

- protect human and animal health and the environment by introducing a safety assessment of the highest possible standards at EU level before any GMO is placed on the market;
- put in place harmonized procedures for risk assessment and authorization of GMOs that are efficient, time-limited and transparent;
- ensure clear labelling of GMOs placed on the market in order to enable consumers as well as professionals (e.g. farmers, and food/feed chain operators) to make an informed choice; and
- ensure the traceability of GMOs placed on the market.

The building blocks of the GMOs legislations include:

- Directive (EU) 2015/412 amending Directive 2001/18/EC as regards the possibility for the Member States to restrict or prohibit the cultivation of GMOs in their territory;
- Directive 2009/41/EC on the contained use of genetically modified microorganisms;
- Regulation (EC) 1830/2003 concerning the traceability and labelling of genetically modified organisms and the traceability of food and feed products produced from genetically modified organisms;
- Regulation (EC) 1829/2003 on genetically modified food and feed;
- Regulation (EC) 1946/2003 on transboundary movements of GMOs; and
- Directive 2001/18/EC on the deliberate release of GMOs into the environment.

These main pieces of legislation are supplemented by a number of implementing rules or by recommendations and guidelines on more specific aspects. The Commission Directive 2018/350 amending Directive 2001/18/EC of the European Parliament and the Council as regards the environmental risk assessment of genetically modified organisms was published on 9 March 2018. This measure brings the requirements on environmental risk assessment (ERA) up to date, with developments in scientific knowledge and technical progress, while building on the EFSA Guidance Document for the ERA of plants. The measure entered into force on 29 March 2018, and the EU countries are requested to bring into force the laws, regulations and administrative provisions necessary to comply with this Directive by 29 September 2019 at the latest. The Commission Implementing Decision 2018/1790 repealing Decision 2002/623/EC was published in the *EU Official Journal* on 20 November 2018. This measure repeals the Guidance Notes of 2002, which have become obsolete. The repeal reduces the number of guidance documents that operators and competent authorities need to take into account when carrying out an environmental risk assessment under Annex II to Directive 2001/18/EC.

8.2 FOOD SUPPLEMENTS

According to the Directive 2002/46/EC, "food supplement" is a food with the objective to integrate the normal diet that is a concentrated source of nutrients (vitamins and minerals) and other substances with a nutritional or physiological effect, by its own or in combination, and is marketed in dosed forms intended to be ingested in small and measured amounts.

Food supplements are marketed as "doses", for example pills, tablets, capsules or liquids in measured quantities to correct nutritional deficiencies or maintain an adequate assumption of specific nutrients. In some cases the excessive assumption of vitamins and minerals may cause undesired effects and it is necessary to establish also maximum levels to ensure safe use of food supplements. The main EU Regulations applicable to food supplements include, in addition to Directive 2002/46/EC:

- Directive 2006/37/EC that amends Annex II of Directive 2002/46/EC with the inclusion of some substances;
- Regulation (EC) 1170/2009 that amends the Directive 2002/46/EC and Regulation (EC) 1925/2006 for the lists of vitamins and minerals and their forms;
- Regulation (EU) 1161/2011 that amends the Directive 2002/46/EC, Regulation (EC) 1925/2006 and that 953/2009 for the lists of mineral substances which can be added to foods; and
- Regulation (EC) 1137/2008 concerning the regulatory procedure of scrutiny.

The vitamins and minerals permitted in food supplements and their molecular forms are those described in Annex I of Directive 2002/46, as amended by Regulation (EC) 1170/2009. The list of permitted bioavailable sources of vitamins and minerals used in food supplements are those reported in Annex II of Directive 2002/46, as modified by Directive 2006/37/EC and by Regulations 1170/2009/EC and 1161/2011/EU.

To food supplements are applicable the purity criteria already specified by Community Regulations for the use of specific substances adopted for other uses. Alternatively, the purity criteria recommended by International Organizations or adopted with national regulations can be utilized. The identification of molecular forms of vitamins and of the food chain minerals that can be used in food supplements has been made possible by EFSA through the evaluation of safety and bioavailability of the nutrient sources proposed by FBOs for their inclusion in Annex II. Between 2005 and 2009, EFSA has examined in total 533 proposals. Of these, 186 were withdrawn during the evaluation process and EFSA received scientific evidences inadequate for evaluating about the half of the remaining applications. Moreover, reasons for safety concern were identified with relation to 39 additional applications.

To define maximum and minimum levels of vitamins and minerals it is important to take into account the elements in the Commission Discussion Paper on the setting of maximum and minimum amounts for vitamins and minerals in foodstuffs (2006):

- the most elevated levels of vitamins and minerals considered safe and the lowest levels considered efficacious on the basis, respectively, of scientific risk evaluation or efficacy, based on generally-accepted scientific data taking into account, as appropriate, the different levels of sensitivity of different groups of citizens;
- the assumption of vitamin and minerals from other dietetic sources; and
- the reference assumptions of vitamins and minerals for the population.

A very relevant issue is that the total amount of each vitamin and mineral present in the food product available on the market should not exceed the respective maximum amount, regardless of the reasons for the addition. However, these maximum levels have not yet been defined by the EC. In such a context it is also important the definition of the "tolerable level of superior assumption" (upper level, UL) that is the maximum daily level of chronic exposure through all sources considered as "unlikely" to induce adverse health effects. EFSA is currently working to this end. As long as the European Commission does not adopt the maximum levels for vitamins and minerals, Member States are authorized to maintain or adopt national levels. As far as the minimum levels of vitamins and minerals are concerned, since 13 December 2014, according to Regulation (EU) 1169/2011 are defined "significant quantities" of vitamins and minerals those corresponding to:

- 15% of nutritional reference values in 100 g or 100 ml for products other than beverages or for a portion if the package contains only one portion; and
- 7.5% of nutritional reference values specified in the previous point for 100 ml in case of beverages.

A report was transmitted by the European Commission to the Council and the Parliament on 5 December 2008 on the use of substances other than vitamins and minerals in food supplements. The number of nutrient substances with nutritional or physiological effects, other than vitamins and minerals used in food supplements, was estimated to be more than 400, belonging to six different groups regulated in different Member States:

- amino acids such as L-arginine and other essential and non-essential amino acids;
- enzymes, such as lactase and papaine;
- prebiotics and probiotics, such as inulin, *Lactobacillus acidophilus*, *Bifidobacterium* species and yeast species;
- essential fatty acids such as gamma-linolenic acid, fish oil (DHA/EPA), oil of *Borago officinalis* and oil of *Linum usitatissimum*;
- botanical products and botanical extracts such as *Aloe vera*, *Ginkgo biloba Panax ginseng*, *Allium sativum*, *Camellia sinensis*, *Garcinia cambogia* and *Paullinia cupana*; and
- other substances such as lycopene, luteine, coenzyme Q10, taurine, carnitine, inositole, glucosamine, chitosan, spirulina and soya isoflavone.

The regulatory status of these substances in the Member States considered indicates that the majority was already permitted in food supplements:

- under national Regulations or internal guidelines;
- below a maximum level or pre-determined specific conditions; and
- on a case-by-case approach following specific evaluations.

Only the minority of these substances was subject to an authorization or was considered to be a medicinal product.

Therefore, the EC concluded that the legal instruments used in the Member States were satisfactory and that there was no need to harmonize these rules for substances other than vitamins and minerals used in food supplements. Therefore, it is a responsibility of competent authorities of Member States to monitor and verify that needed requirements in terms of safety and efficacy are satisfied. Moreover, the Regulation on the free circulation of safe products and the procedure of mutual recognition is also applicable to food supplements (see Articles 11 and 12 of Directive 2002/46/EC). To facilitate the knowledge of the food supplements market, Member States may adopt the procedure of label notification for the new supplements before their admission in the national market. In relation to the evaluation of the safety of food products:

- for food supplements characterized by a significant use before May 1997, a presumption of safety is applied in the absence of adverse effects associated with the historical use; and
- for food supplements without any significant use before May 1997, the EU Regulation on novel foods is to be applied (see Section 8.3 in the present book).

A guide on the evaluation of the safety of botanical parts and preparations intended for use for manufacturing food supplements has been produced by EFSA (*EFSA Journal* 2009; 7(9): 1249). This methodology is based essentially on the acceptance of exposure levels corresponding to the range derived from historical food use in the EU without any evidence of adverse health effects in case of no substances of concern have been identified. A specific compendium (database) of substances of concern occurring in different botanical species is currently being developed by EFSA (see Section 11.7.1).

The Member States cannot prohibit or limit the marketing of food supplements which are in compliance with the EU Regulations. In case a specific product represents a danger for public health, the Member State can temporarily suspend or limit the use of the product considered unsafe, informing at the same time the other Member States and the European Commission on the reasons motivating the action. The EC evaluates the motivations offered by the Member State to suspend or limit temporarily the marketing of the food supplement and, after having consulted the Permanent Committee for the Food Chain and Animal Health, adopts the measures considered necessary.

The consumer information on food supplements must be in compliance with three Regulations:

- Directive 2002/46/CE;
- Regulation (EU) 1169/2011; and
- Regulation (EC) 1924/2006.

According to Regulation (EU) 1169/2011, in the labelling of food supplements it is mandatory:

- the name under which the product is sold;
- the list of ingredients;
- the quantity of certain ingredients or categories of ingredients;
- the net quantity;
- the date of minimum durability or, in the case of foodstuffs which, from the microbiological point of view, are highly perishable, the "use by" date;

121

- any special storage conditions or conditions of use;
- the name or business name and address of the manufacturer or packager, or of a seller established within the Community;
- particulars of the place of origin or provenance where a failure to give such particulars might mislead the consumer to a material degree as to the true origin or provenance of the foodstuff;
- instructions for use when it would be impossible to make appropriate use of the foodstuff in the absence of such instructions; and
- with respect to beverages containing more than 1.2% by volume of alcohol, the actual alcoholic strength by volume.

The obligation to provide the nutritional declaration does not apply to food supplements. The Directive 2002/46/EC provides several mandatory labelling rules which include:

- the name "food supplement";
- the names of the categories of nutrients or substances that characterize the product or an indication of the nature of these nutrients and substances;
- the portion of the product recommended for the daily consumption;
- a warning to not exceed the recommended daily acceptable dose;
- a warning to not use the food supplement as a substitute for a varied diet;
- a warning to inform that the product should be kept out children hands;
- the amounts of nutrients or of substances with a nutritional or physiological effect present in the product must be declared on the label in numerical forms;
- the amounts of nutrients or of substances with a nutritional or physiological effect must be those per portion of the product recommended for daily consumption; and
- the information on vitamins and minerals shall also be expressed as percentages of the reference nutritional values reported in Annex XIII of Regulation (EU) 1169/2011.

The Directive 2002/46/CE also provides several prohibitions such as:

- do not attribute properties able to prevent, treat or care human diseases to food supplements; and

- the labelling, the presentation and the publicity of food supplements should not include any citation that states or implies that a balanced and varied diet may not provide in general adequate amounts of nutrients.

As shown by the example of trehalose (Box 8.1), with relation to the labelling of new ingredients in food supplements, there are specific additional requirements. When necessary, it is possible to mention:

- the characteristics: composition, nutritional value and use;
- materials which may affect the health of specific individuals; and
- materials which may generate ethic concerns.

BOX 8.1   THE EXAMPLE OF TREHALOSE

The designation "trehalose" shall be displayed on the labelling of the product as such or in the list of ingredients of foodstuffs containing it. In a prominently displayed footnote related to the designation "trehalose" by means of an asterisk (*) the words "trehalose is a source of glucose" shall be displayed. The words shall have a typeface of at least the same size as the list of ingredients itself.

Source: Commission Decision 2001/721/EC

To food supplements it is also possible to apply Regulation (EC) 1924/2006 on voluntary labelling. However, in the case of food supplements, nutritional claims on ingredients whose intakes should be reduced (e.g. saturated fats, sugars and salt) and comparative claims do not find applications. In the case of health claims, considerable difficulties have emerged for specific food supplements such as probiotics and other products. For probiotics it has been so far impossible to authorize any health claims although as many as about 300 applications were presented to EFSA, due to:

- poor products characterization;
- non-specific or non-beneficial physiological effects;
- difficulties in connecting the strains used for the studies with those for which the claims were proposed; and
- limited human studies.

Following this first evaluation and the proposal by EFSA of a methodology for an adequate characterization of probiotics, for a considerable number of applications additional data were presented to make possible a new evaluation by EFSA. Such a development made possible for EFSA to recognize as satisfactory the characterization of products for a large number of applications supported with additional data. However, all the applications concerning claims on probiotics have received a negative evaluation by EFSA for one or more of the previously mentioned reasons. Moreover, about 2,000 applications for health claims already in use on botanical food supplements were presented to the EC and EFSA in the framework of the procedure of Article 13(1) of Regulation (EU) 1924/2006. Most of these applications did not receive a positive evaluation by EFSA due to an insufficient characterization or because a cause and effect relation had not established. On 20 October 2010, the EC asked EFSA to discontinue the evaluation of botanical food supplements and the claims in question were allowed to remain on the market.

8.3 FORTIFIED FOODS

The EU Regulations applicable to fortified foods are:

- Regulation (EC) 1925/2006 on the addition of vitamins and minerals and of certain other substances to foods;
- Regulation (EC) 108/2008 amending Regulation (EC) 1925/2006;
- Commission Regulations EC/1170/2009 and EU/1161/ 2011 which amended Annex I and Annex II to include additional vitamin or mineral formulations;
- Commission Implementing Regulation (EU) 307/2012 establishing implementing rules for the application of Article 8 of Regulation (EC) 1925/2006; and
- Commission Implementing Regulation (EU) 489/2012 establishing implementing rules for the application of Article 16 of Regulation (EC) 1925/2006.

The objectives of these Regulations include:

- the harmonization of the Member States Regulations concerning the addition of vitamins, minerals and some other substances to foods;
- ensuring the effective functioning of the internal market that makes possible equal conditions of competition; and
- ensuring higher levels of consumer's protection.

The food fortification aims at avoiding or reducing the states of deficiency concerning vitamins and minerals caused by a lack of foods mainly contributing to these nutrients or by very unbalanced diets. A specific type of food fortification is defined as "restoration". In the restoration, the final product recovers nutrient amounts lost during the different steps of food conservation, manipulation and transformation. In the "fortification", vitamins and minerals are added to foods independently on whether they were originally present or not in the food. Vitamins and minerals are those listed in Annex I of Regulation (EC) 1925/2006, as amended from Regulations EC/108/2008, EC/1170/2009 and EU/1161/2011. The permitted bioavailable vitamin and mineral forms are those listed in Annex II as amended by Regulations EC/108/2008, EC/1170/2009 and EU/1161/2011.

Vitamins and minerals may be added to foods, taking into account:

- the lack of one or more vitamins or minerals for all the population or in a specific group of the population;
- the potential improvement of the nutritional state that can be achieved through the diet supplementation with specific vitamins and/or minerals; and
- the evolution of scientific knowledge on the roles of vitamins and minerals on nutrition and health.

Annexes I and II of Regulation (EC) 1925/2006 can be updated with the comitology procedure. It is not possible to add vitamins and minerals to non-transformed foods such as fruits, vegetables, meat, chicken and fish as well as to beverages containing more than 1.2% by volume alcohol (exceptions: "tonic wine" or "ginger tonic wine" notified by Ireland and the UK for a derogation).

For vitamins and minerals in fortified foods it is possible to apply the purity criteria already specified by Community Regulations for specific substances adopted for other uses. Alternatively, the purity criteria recommended by International Organizations or adopted with national Regulations can be utilized. To define maximum and minimum levels of vitamins and minerals, it is important to take into account the elements discussed in the Commission Discussion Paper on the setting of maximum and minimum amounts for vitamins and minerals in foodstuffs (2006):

- the most elevated levels of vitamins and minerals considered safe and the lowest levels considered efficacious on the basis, respectively, of scientific risk evaluation or efficacy, based on generally-accepted scientific data by taking into account, as appropriate, the different levels of sensitivity of different groups of citizens;

- the assumption of vitamins and minerals from other dietetic sources; and
- the reference assumptions of vitamins and minerals for the population.

The criteria for the choice of maximum and minimum levels of vitamins and minerals in fortified foods are essentially the same described in the previous chapter for food supplements.

Fortified foods cannot be put on the market if they are not safe because dangerous for human health or non-adequate for human consumption. According to Article 17 of Regulation (EC) 178/2002: food business operators, in all the steps of the food chain under their control, should ensure and verify that all foods satisfy all the requirements of the food legislation. In case of potential risks deriving from the addition to foods of other substances, the procedure of Article 8 of Regulation (EC) 1925/2006 should be applied. The Article 8 procedure shall be followed in case the consumption of the fortified food would:

- cause the ingestion of amounts of a specific substance largely exceeding those considered normal in a balanced and varied diet; and/or
- represent a potential risk for the consumer (general adult population or other specific groups of the population).

Such conditions should be verified under use circumstances and evaluated through a case-by-case approach in relation to the average assumption of the interested substances from the adult population or specific population groups. The substances listed in Annex III of Regulation EC/1925/2006 may be prohibited, restricted or under Community scrutiny:

- PART A – the substance has a dangerous effect and its addition to food is prohibited;
- PART B – the substance has a dangerous effect and its addition to food may be done only under specific limitations; and
- PART C – a possible dangerous effect has been identified, but due to persisting uncertainties a four-year scrutiny is ongoing.

A Community Register is foreseen by the Regulation on fortified food to report:

- Annexes I, II and III;
- maximum and minimum amounts of vitamins and minerals;
- national rules concerning mandatory addition of vitamins and minerals;
- current restrictions to the addition of vitamins and minerals and of other substances; and
- lists of vitamin and mineral forms for which dossiers have been presented.

The nutritional declaration is mandatory for foods added with vitamins and minerals.

8.4 NOVEL FOODS

Regulation (EU) 2015/2283 on novel foods is applicable since 1 January 2018. It amends Regulation (EU) 1169/2011 of the European Parliament and the Council and repeals Regulation (EC) 258/97 of the European Parliament and the Council and Commission Regulation (EC) 1852/2001, which were in force until 31 December 2017, leading to the entry in the European market of a considerable number of novel foods (Paoletti et al., 2011a).

This Regulation has the objective of improving conditions so that food businesses can easily bring new and innovative foods to the EU market, while maintaining a high level of food safety for European consumers.

The main features and improvements of this new Regulation are the following (Silano and Carnassale, 2018):

- **Expanded categories of novel foods**: The "novel food" definition describes the various situations of foods originating from plants, animals, microorganisms, cell cultures, minerals, specific categories of foods (insects, vitamins, minerals, food supplements, etc.), foods resulting from production processes and practices, and state-of-the-art technologies (e.g. intentionally modified or new molecular structure, nanomaterials), which were not produced or used before 1997 and thus may be considered to be as novel foods.
- **Generic authorizations of novel foods**: under the new Regulation, all authorizations (new and old) are generic as opposed to the applicant-specific, restricted novel food authorizations under the old "novel food" regime. This means that any FBO can place an

authorized novel food on the European Union market, provided the authorized conditions of use, labelling requirements and specifications are respected.

- **Establishment of a Union list of authorized novel foods**: This is a positive list containing all authorized novel foods. Novel foods which will be authorized in the future will be added to the Union list by means of Commission Implementing Regulations. Once a novel food is added to the Union list, then it is automatically considered as being authorized, and it can be placed in the European Union market.
- A simplified, centralized authorization procedure managed by the European Commission using an online application submission system.
- Centralized, safety evaluation of the novel foods will be carried out by EFSA. The European Commission consults EFSA on the applications and bases its authorization decisions on the outcome of the EFSA's evaluation.
- Efficiency and transparency will be improved by establishing deadlines for the safety evaluation and authorization procedure, thus reducing the overall time spent on approvals.
- A faster and structured notification system for traditional foods from third countries on the basis of a history of safe food use. To facilitate the marketing of traditional foods from countries outside the EU, which are considered novel foods in the EU, the new Regulation introduces a simplified assessment procedure for foods new to the EU. If the safety of the traditional food in question can be established on the basis of evidence of a history of consumption in a third country, and there are no safety concerns raised by the EU countries or EFSA, the traditional food will be allowed to be placed on the European Union market.
- Promotion of innovation by granting an individual authorization for five years based on protected data. Data protection provisions are included in the new Regulation. That means that an applicant may be granted an individual authorization for placing on the market of a novel food. This is based on newly developed scientific evidence and proprietary data and is limited in time to five years.

Article 4 of the novel food Regulation (EU) 2015/2283 requires FBOs to verify if the food they intend to place on the EU market falls within the scope of the novel food regulation, that is whether the food is novel or not. If, after considering all the information available, food business operators are still unsure about a food as being novel, they may consult the competent authorities of the EU country where they first intend to place the food (called "the recipient EU country") on the market. Commission Implementing Regulation (EU) 2018/456 lays down the information requirements that need to be included in the consultation request, including provisions on the confidentiality of the request, and the procedural steps FBOs must follow for the consultation process. Once the recipient EU country reaches its conclusion on the "novel status" of a food, the Commission will publish that information on the Commission's website. To facilitate the entry into force of the Regulation on novel foods, the European Commission has adopted the Commission Implementing Regulation (EU) 2017/2469 laying down administrative and scientific requirements for applications referred to in Article 10 of Regulation (EU) 2015/2283 of the European Parliament and the Council on novel foods. This implementing act sets out the administrative, technical and scientific requirements which should be included in a novel food application. In addition, specifically on the scientific requirements for novel foods, EFSA has also issued a detailed guidance. For novel foods, EFSA carries out its safety assessment based on dossiers provided by applicants. These dossiers need to contain data on the compositional, nutritional, toxicological and allergenic properties of the novel food as well as information on respective production processes and the proposed uses and use levels (see the EFSA Guidance). For traditional foods from third countries that may constitute a subset of novel foods, EFSA, in parallel with the Member States, assesses the safe use based on information provided by the applicant. This evidence has to demonstrate the safe use of the traditional food in at least one country outside the EU for a period of at least 25 years (see the EFSA Guidance). EFSA does not decide whether a food is considered a novel food or a traditional food from a third country but only provides a scientific evaluation of the available safety data; the final decision on the novelty of the food is the responsibility of EU risk managers. Similarly, risk managers decide whether a novel food or a traditional food from a third country can be placed on the EU market, including its authorized conditions of use. Both the legal text of the implementing act and the EFSA Guidance documents are intended to assist food business operators and facilitate the preparation and submission of a novel food application.

Other relevant implementing Regulations to be carefully considered in this context are the following:

- Commission Implementing Regulation (EU) 2017/2468 laying down administrative and scientific requirements concerning traditional foods from third countries in accordance with Regulation (EU) 2015/2283 of the European Parliament and of the Council on novel foods. This implementing act sets out the administrative, technical and scientific requirements which should be included in a notification of a traditional food from a third country and which is considered novel in the European Union; and
- Commission Implementing Regulation (EU) 2017/2470 on establishing the Union list of novel foods in accordance with Regulation (EU) 2015/2283 of the European Parliament and the Council on Novel Foods.

In an article published (Lahteenmaki et al, 2016), an analysis is carried out to evaluate whether the new novel foods Regulation is helpful for promoting the use of insects, considered to have a high potential as a novel food and feed. The overall conclusion has been that, even though the new novel food Regulation has finally clearly included whole insects under the scope of its application, the EU still needs to solve a number of outstanding legal questions in order to promote innovation and growth while guaranteeing food and feed safety such as a Regulation on living and killing conditions of insects and the requirements or the hygienic production of insect-based food and feed, considering that insects can also eat by-products and wastes. For a very similar analysis with related but also different conclusions, see also the article published by Finardi and Derrien (2016). The new novel food Regulation does not apply to the following food products and the Regulations listed below will continue to apply:

- food enzymes within Regulation (EC) 1332/2008;
- food additives within Regulation (EC) 1333/2008;
- flavourings for use in foods within Regulation (EC) 1334/2008;
- extraction solvents used in the production of foods within Directive 2009/32/EC approximating EU countries' laws; and
- GMOs for food and feed, covered by Regulation (EC) 1829/2003.

If foods and/or food ingredients were used exclusively in food supplements, new uses in other foods require authorization under the novel food Regulation.

9

Foods for Specific Population Groups

9.1 INTRODUCTION

Regulation (EU) 609/2013 of the European Parliament and of the Council of 12 June 2013 on food intended for infants and young children, food for special medical purposes, and total diet replacement for weight control ("Food for Specific Groups") and repealing Council Directive 92/52/EEC, Commission Directives 96/8/EC, 1999/21/EC, 2006/125/EC and 2006/141/EC, Directive 2009/39/EC of the European Parliament and of the Council and Commission Regulation (EC) 41/2009 and (EC) 953/2009, applied since 20 July 2016. It aims at protecting specific vulnerable groups of consumers by regulating the content and marketing of food products specifically created for and marketed to them. It also aims to increase legal clarity for business and to facilitate the correct application of the rules.

This Regulation:

- sets general compositional and labelling rules and require the Commission to adopt, through delegated acts, specific compositional and labelling rules for:
  - infant and follow-on formula;
  - processed cereal-based food and other baby food;
  - food for special medical purposes; and
  - total diet replacement for weight control.
- simplifies the regulatory framework by eliminating unnecessary and contradictory rules and by replacing them with a new

framework which takes into account the developments on the market and in EU food law. In particular, this Regulation abolishes the obsolete concept of "dietetic food" by repealing Directive 2009/39/EC, which previously laid down general rules for these products categories (specific rules on the different product categories adopted in the past under Directive 2009/39/EC have remained applicable until the adoption of the new specific rules);

- establishes a single Union list of substances that can be added to these foods including minerals and vitamins;
- empowers the European Commission to adopt interpretation decisions clarifying whether a given food falls within the scope of the Regulation and under what specific food category, in order to ensure uniform implementation of the rules;
- requires the European Commission to transfer rules on gluten-free foods and very-low gluten under Regulation (EU) 1169/2011 on food information to consumers in order to ensure clarity and consistency; and
- establishes that meal replacement products for weight control should be regulated solely under Regulation (EC) 1924/2006 on nutrition and health claims in order to ensure legal certainty.

Moreover, this Regulation foresees that two reports should be prepared by the European Commission in order to analyze the need to establish special rules for:

- young-child formulae (the so-called growing-up milks); and
- food intended for sportspeople (European Commission, 2016a,b).

The report on young-child formulae was adopted by the European Commission in 2016, together with a document with more detailed information on its findings. The report on food intended for sportspeople was adopted by the European Commission on 15 June 2016. This report is mainly based on the results of an external study commissioned by DG HEALTH to the Food Chain Evaluation Consortium (FCEC) titled "Study on Food Intended for Sportspeople". Both reports concluded that there is no need for specific provisions for these products. From 20 July 2016, young-child formulae and food intended for sportspeople are exclusively covered by horizontal rules of EU food law. A similar conclusion was reached on foods for diabetics. In 2008, the European Commission produced a report on the desirability of special provisions for foods intended for people suffering from diabetes. Diabetes is a metabolic disorder due to impaired insulin secretion by the pancreas or reduced sensitivity of insulin receptors on the cellular membrane of hepatocytes and muscle cells. The report was required by the old legislative framework of Directive 2009/39/EC. This report concluded that there was no scientific basis to produce specific EU legislation covering compositional requirements for this group of foods. Following the report's conclusions, the European Parliament and Council agreed to exclude foods for diabetics from the scope of Regulation (EU) 609/2013 on foods for specific groups. The report from the Commission to the European Parliament and the Council on foods for persons suffering from carbohydrate metabolism disorders (diabetes) was adopted in 2013.

9.2 FOODS FOR SPECIAL MEDICAL PURPOSES

The rules for the composition and labelling of foods intended for the dietary management (under medical supervision) of individuals who suffer from certain diseases, disorders or medical conditions have been laid down by the Commission Directive 1999/21/EC. These foods are intended for the exclusive or partial feeding of people whose nutritional requirements cannot be met by normal foods (e.g. inherited metabolic diseases, malnutrition, cystic fibrosis, dysphagia, pre-term formulae and short intestine). The Directive 1999/21/EC lays down essential requirements for their composition and gives guidance for the minimum and maximum levels of vitamins and minerals. Nutritional substances that may be used in the manufacture of foods for special medical purposes are laid down in Commission Regulation (EC) 953/2009. For foods for special medical purposes, the Regulation (EU) 609/2013 on "Food for Specific Groups" has from 20 July 2016:

- set general compositional and labelling rules; and
- required the Commission to adopt, through a delegated act, specific compositional and labelling rules for foods for special medical purposes to replace Directive 1999/21/EC.

The Commission delegated Regulation (EU) 2016/128 was adopted on 25 September 2015 and is applied since 22 February 2019, while the Directive 1999/21/EC has remained applicable in the meantime.

The Regulation (EU) 2016/128:

- maintains the rules of Directive 1999/21/EC with some changes to the labelling requirements to ensure consistency with horizontal rules of Regulation (EU) 1169/2011 on the provision of food information to consumers, taking into account the specificities of the products;

- introduces the prohibition to make nutrition and health claims on foods for special medical purposes, in order to ensure legal clarity and avoid inappropriate promotion of the products;
- extends to foods for special medical purposes intended for infants, all rules on labelling, presentation, advertising and marketing applicable to infant formulae for healthy infants that would not be contrary to the products' intended use. This will ensure consistency of EU rules and contribute to avoiding misclassification of products; and
- extends to foods for special medical purposes intended for infants and young children the same rules on pesticides that apply to infant formula, follow-on formula, processed cereal-based foods and baby foods.

9.3 FOODS FOR INFANTS AND YOUNG CHILDREN

Infant formulae (intended as the sole nutritional source for infants from 0 to 6 months, when breastfeeding is not available or insufficient) and follow-on formulae (intended as a nutritional source, complemented with other liquid or solid foods, for infants from 6, or 4 under the paediatrician advice, to 12 months, when breastfeeding is not available or insufficient) are products designed to satisfy the specific nutritional requirements of healthy infants. These products are currently specifically covered by Commission Directive 2006/141/EC, amended on several occasions, that established the requirements for the composition and labelling of infant formulae and follow-on formulae. The annexes of the Directive gave criteria for the composition (energy, protein, carbohydrate, fat, mineral substances, vitamins and certain other ingredients) of infant formulae and follow-on formulae including, where necessary, minimum and maximum levels. Commission Regulation (EC) 1609/2006 authorized the placing on the market of infant formulae based on hydrolysates of cow's milk in accordance with specifications for the protein content source, processing and quality set out in an Annex for a period of two years. Further to the expiry of such Regulation, these specifications concerning protein quality were added to Directive 2006/141/EC by adopting Commission Regulation (EC) 1243/2008. Commission Directive 2013/46/EU extended these specifications to follow-on formulae based on hydrolysates of cow's milk and also authorized the placing on the market in the EU of infant formulae and follow-on formulae manufactured from goats' milk protein. Directive 2006/141/EC also encompassed the specific rules on the presence of pesticide residues in infant and follow-on formulae, previously set out in Commission Directive 1999/50/EC. It is required that infant formula and follow-on formula contain no detectable levels of pesticide residues, meaning no more than 0.01 mg of pesticide residues per kg. The Directive also prohibited the use of certain toxic pesticides in the production of infant and follow-on formulae and established levels lower than the general maximum level of 0.01 mg per kg for a few other very toxic pesticides.

In addition to the requirements in Directive 2006/141/EC, infant formulae and follow-on formulae must also comply with other specific provisions laid down in the relevant measures of EU law on hygiene, on the use of food additives, on the presence of contaminants and on the use of materials intended to come into contact with the products. The Regulation (EU) 2013/609 on "Food for Specific Groups" has from 20 July 2016:

- set general compositional and labelling rules. In addition, it extends to the labelling of follow-on formula the restriction of the use of pictures or text which may idealize the use of products (previously only applicable to infant formula); and
- required the European Commission to adopt, through a delegated act, specific compositional and information rules for infant and follow-on formulae, to replace Directive 2006/141/EC. Commission delegated Regulation (EU) 2016/127 was adopted on 25 September 2015 and started to apply on 22 February 2020. Until that date, the rules of Directive 2006/141/EC have remained in force.

The new delegated Regulation:

- updates the existing compositional requirements on the basis of the latest scientific advice;
- modifies the rules on labelling to ensure consistency with horizontal rules of Regulation (EU) 1169/2011 on the provision of food information to consumers, taking into account the specificities of the products;
- forbids the use of nutrition and health claims on infant formula to protect breastfeeding; and

- facilitates the monitoring activities of competent national authorities by requiring operators to notify them of the placing on the market of many follow-on formulae (in addition to infant formulae, for which the obligation already existed).

From 22 February 2021, infant formula and follow-on formula manufactured from protein hydrolysates will have to comply with the new requirements of the delegated Regulation (EU) 2016/127. As explained in the recitals, these requirements may be updated in the future in order to allow the placing on the market of formulae manufactured from protein hydrolysates with different compositions following a case-by-case evaluation of their safety and suitability by EFSA. In addition, after the EFSA assessment, on the basis of studies, where it is demonstrated that a specific formula manufactured from protein hydrolysates reduces the risk of developing an allergy to milk proteins, further consideration will be given to how adequately inform parents and caregivers about that property of the product. The "Administrative Guidance on the Submission of Dossiers on Infant and/or Follow-on Formula Manufactured from Protein Hydrolysates" provides detailed information on the procedure that should be followed for the submission of dossiers related to infant and/or follow-on formula manufactured from protein hydrolysates in the context of delegated Regulation (EU) 2016/127.

9.4 CEREALS-BASED FOODS AND OTHER BABY FOODS

Processed cereal-based foods and other baby foods (*weaning foods*) are specifically intended for infants (children from 4 months and under the age of 12 months) and young children (between one and three years) as they progress onto a mixed family diet. Processed cereal-based foods and baby foods for infants and young children are currently covered by Commission Directive 2006/125/EC, adopted under the old legislative framework. It sets out rules on the composition and labelling of processed cereal-based foods and other baby foods. It gives criteria for the composition (protein, carbohydrate, fat, mineral substances and vitamins) of weaning foods including, where necessary, minimum and maximum levels. The Directive encompasses the specific rules on the presence of pesticides residues in processed cereal-based baby foods and baby foods set out in Commission Directive 99/39/EC and requires that this type of food contains no detectable levels of pesticide residues, meaning no more than

136

0.01 mg of pesticide residues per kg. In addition, the Directive prohibits the use of certain very toxic pesticides in the production of processed cereal-based baby foods and baby foods and establishes levels lower than the general maximum level of 0.01 mg per kg for a few other very toxic pesticides.

In addition to the requirements in Directive 2006/125/EC, processed cereal-based foods and baby foods must also comply with other specific provisions laid down in the relevant measures of EU law on hygiene, on the use of food additives, on the presence of contaminants and on the use of materials intended to come into contact with the products.

The Directive (EC) 125/2006:

- sets general compositional and labelling rules; and
- requires the Commission to adopt, through a delegated act, specific compositional and labelling rules for processed cereal-based foods and baby foods, which will replace Directive 2006/125/EC. Until the finalization of the delegated act, the rules of Directive 2006/125/EC remain in force.

9.5 FOODS FOR WEIGHT REDUCTION (CONTROL) AND COMPOSITIONAL REQUIREMENTS FOR TOTAL DIET REPLACEMENT FOR WEIGHT CONTROL PRODUCTS

Foods for weight control are divided into two categories:

(a) products presented as a replacement for the whole of the daily diet (800–1,200 kcal per total daily ration); and
(b) products presented as a replacement for one or more meals of the daily diet (200–400 kcal per meal).

Commission Directive 96/8/EC of 26 February 1996 currently lays down compositional and labelling requirements for foods intended to be used in energy-restricted diets for weight reduction. The compositional criteria include requirements on energy, protein quantity and quality, fat quantity and type, minimum and maximum levels for dietary fibre and minimum levels for certain vitamins and minerals. Nutritional substances that may be used in the manufacture of these products are laid down in Commission Regulation (EC) 953/2009. Commission Directive 2007/29/ EC of 30 May 2007 has amended Directive 96/8/EC to adapt its rules to Regulation (EC) 1924/2006 on nutrition and health claims made on foods.

Through this Directive it is also allowed to make claims referring to a reduction in the sense of hunger or an increase in the sense of satiety if the claims comply with EU legislation.

For foods for weight reduction, the Regulation (EU) 609/2013 on Food for Specific Groups has:

- set general compositional and labelling rules for total diet replacement products for weight control;
- required the European Commission to adopt, through a delegated act, specific compositional and labelling rules for total diet replacement products for weight control. Until the finalization of the delegated act, the rules of Directive 96/8/EC remain applicable to these products; and
- established that rules on the use of statements on meal replacement products (between 840 kJ (200 kcal) and 1,680 kJ (400 kcal)) should be regulated solely under Regulation (EC) 1924/2006 on nutrition and health claims in order to ensure legal certainty.

The delegated Regulation (EU) 2017/1798 on compositional requirements for the total diet replacement for weight control products was adopted on 2 June 2017 and is applicable since 27 September 2022. This Regulation has determined the following requirements with regard to "total diet replacement for weight control" products:

1. Compositional requirements as listed in Annex 1 for energy, protein, choline, lipids (linoleic and alpha-linolenic acid), carbohydrates and vitamins and minerals and in Annex 2 for selected amino acids.
2. Requirements for labelling, presentation and advertising as specified below:
   (a) a statement that the product is only intended for healthy overweight or obese adults who intend to achieve weight reduction;
   (b) a statement that the product should not be used by pregnant or lactating women, adolescents or by individuals suffering from a medical condition without the advice of a healthcare professional;
   (c) a statement on the importance of maintaining an adequate daily fluid intake;
   (d) a statement that the product provides adequate daily amounts of all essential nutrients when used in accordance with the instructions for use;

(e) a statement that the product should not be used for more than 8 weeks, or repeatedly for shorter periods, by healthy over-weight or obese adults without the advice of a healthcare professional;

(f) instructions for appropriate preparation, where necessary, and a statement as to the importance of following those instructions;

(g) if a product, when used as instructed by the manufacturer, provides a daily intake of polyols in excess of 20 g per day, a statement that the food may have a laxative effect; and

(h) if dietary fibre is not added to the product, a statement that the advice of a healthcare professional must be sought regarding the possibility of supplementing the product with dietary fibre.

When appearing on the package or on the label attached thereto, the mandatory particulars listed in paragraph 1 shall be indicated in such a way as to meet requirements laid down in Article 13(2) and (3) of Regulation (EU) 1169/2011. The labelling, presentation and advertising of total diet replacement for weight control products shall not make any reference to the rate or amount of weight reduction, which may result from its use.

3. Specific requirements concerning the nutrition declaration as detailed in Article 5.

Nutrition and health claims shall not be made on the total diet replacement for weight control products, with the only exception that the nutrition claim "added fibre" may be used for the total diet replacement for weight control products provided that the dietary fibre content of the product is not less than 10 g. When the total diet replacement for weight control products are placed on the market, the food business operator shall notify the competent authority of each Member State where the product concerned is being marketed of the information appearing on the label, by sending a model of the label used for the product and of any other information the competent authority may reasonably request to establish compliance with this Regulation, unless a Member State exempts the food business operator from that obligation under a national system that guarantees efficient official monitoring of the product concerned. References to Directive 96/8/EC in other acts shall be construed as references to this Regulation. Given the important changes introduced in total diet replacement for weight control products, particularly with regard to the increase in the minimum amount of protein and essential fatty acids, this legislation has granted a significant period for adaptation to the new Regulation. Moreover, nutrition and health claims shall not be permitted in total diet replacement for weight control products and for these products it will not be possible to refer to the speed or magnitude of weight loss achievable through their use. However, some specific declarations (e.g. "low-calorie diet" or "added fibre") may be allowed, provided that the conditions laid down by the Regulation are met (Banares Viella and Vaqué, 2018).

9.6 GLUTEN-FREE FOODS

Celiac Disease (CD) is a permanent autoimmune enteropathy, triggered in susceptible individuals, by the ingestion of gluten, a storage protein fraction present in wheat grain, and of similar proteins of rye and barley. Approximately 96% of CD patients express the HLA molecule DQ2, while the remainder mostly express the less common haplotype DQ8. All the intestinal gluten-sensitive T cell clones are DQ restricted, reflecting the critical role of this molecule in the pathogenesis of CD. After being bound to DQ, certain deamidated gluten peptides are recognized by interferon-g (IFN-gamma) producing T-cells, thus driving the inflammatory response that causes the histological features of celiac intestinal mucosa. The only known treatment of CD is the life-long withdrawal of gluten-containing food from the diet. Complying with a gluten-free diet is difficult because of the wide distribution and consumption of cereal-based foods, but strict adherence is necessary to reduce mortality and morbidity.

A gluten-free diet encompasses naturally gluten-free food (meat, fish, eggs, pulses, vegetables and fruit) and industrially transformed food products where the gluten content is kept below the 20 ppm by means of physical and chemical gluten degradation or using naturally gluten-free cereals (e.g. maize and rice). Commission Implementing Regulation (EU) 828/2014 lays down harmonized requirements for the provision of information to consumers on the absence or reduced presence of gluten in food. More specifically, this legislation sets out the conditions under which foods may be labelled as "gluten-free" or "very-low gluten". The Implementing Regulation entered into application on 20 July 2016. On the same day, Commission Regulation (EC) 41/2009 concerning the composition and labelling of foodstuffs suitable for people intolerant to gluten, adopted under the old legislative framework of Directive 2009/39/EC, was repealed.

Regulation (EU) 609/2013 required the Commission to transfer its rules under the framework of Regulation (EU) 1169/2011 on the provision of food information to consumers. Regulation (EU) 1169/2011 lays down rules requiring the mandatory labelling for all foods of ingredients such as gluten-containing cereals, with a scientifically proven allergenic effect. In order to ensure clarity and consistency, it was considered that all the rules applying to gluten should be set by the same piece of legislation, and, for this reason, Regulation (EU) 609/2013 established that Regulation (EU) 1169/2011 should also be the framework for the rules related to information on the absence of gluten in food. In order to comply with the requirements of Regulation (EU) 609/2013, the European Commission:

- first amended Regulation (EU) 1169/2011 through Commission delegated Regulation (EU) 1155/2013 (this allowed the Commission to lay down rules on the matter); and
- subsequently laid down the specific requirements in Commission Implementing Regulation (EU) 828/2014 that did not change the substantial rules for using the "gluten-free" and "very-low-gluten" statements previously laid down in Regulation (EC) 41/2009. However, the new rules apply also to pre-packed foods such as those served in restaurants that were out of the scope of the old rules. In addition, the new Regulation also clarifies how operators can inform celiac consumers of the difference between foods that are naturally free of gluten and products that are specially formulated for them. Some EU Member States National Health Services support the gluten-free food provision to CD patients in variable monthly amounts. For additional information on the absence or reduced presence of gluten in food see the article published by Vaquè (2014).

10

Regulations on Animal and Plant Health and on Organic Food and Country of Origin or Place of Provenance

10.1 INTRODUCTION

Relevant Regulations in the area of food and feed safety include those on animal health and welfare and plant health.

Main animal health issues currently include:

(i) African swine fever;
(ii) avian influenza;
(iii) *Echinococcus multilocularis*; and
(iv) lumpy skin diseases,

whereas the main current animal welfare issues include:

(i) rabbit production;
(ii) slaughter of animals; and
(iii) killing of animals for purposes other than for slaughter.

In the area of plant health the emphasis is currently on the control of the introduction and spread of plant pests, such as fungi, bacteria, viruses and insects, among food crops, natural vegetation and landscape plants, which is a serious threat that can have far-reaching economic, social and environmental consequences.

10.2 ANIMAL HEALTH

Animal health has received considerable attention from the European Union Institutions (Belluzzi et al., 2014). As clearly stressed in the European Commission website, in 2007 the EU adopted an "animal health strategy" focused on preventive measures, disease surveillance, controls and research to reduce the incidence of potentially devastating animal diseases and minimize the impact of outbreaks. The main goal was to welcome a new EU animal health policy that would be more robust but flexible, efficient and effective. The strategy had four pillars, along with two key underlying principles consisting of partnership and communication:

- defining priorities with interventions and resources focused on diseases with high public and especially EU relevance, in line with the principles of proportionality and subsidiarity;
- implementation of the EU animal Health Framework The main objective of the EU animal heath framework is the development of an EU Animal Health Law to replace the existing series of linked and interrelated policy actions with a single policy framework to become the ultimate legal vehicle for implementing many principles of the strategy;
- prevention, surveillance and preparedness to identify problems before they emerge while being ready to manage outbreaks and crises; and
- science, innovation and research to ensure adequate responses and maintain a high level of consumer confidence and trust across the EU.

In September 2008, the Commission adopted an action plan to facilitate a better implementation of the strategy, based on the:

- Animal Health Advisory Committee;
- more effective communication and partnership with stakeholders;
- development of a new EU Regulation on animal health;
- revision of the rules on animal by-products and EU expenditures;
- signature of a Memorandum of Understanding with OIE;

- increase of monitoring and surveillance of animal diseases across the EU;
- introduction of bovine electronic identification;
- reinforcement of animal disease;
- emergency preparedness; and
- improvement of import controls on animals and animal products.

The Animal Health Law was adopted with Regulation (EU) 2016/429 of the European Parliament and the Council on 9 March 2016 that was published in the *EU Official Journal* on 31 March 2016. The Regulation entered into force on the 20th day following its publication in *Official Journal of the European Union* and is expected to be fully applicable in five years (see the European Commission website). Overall, the new Animal Health Law supports the EU livestock sector in its quest towards competitiveness and safe and smooth EU market of animals and their products, leading to growth and jobs in this important sector through:

- streamlining a huge number of legal acts into a single law;
- adopting simpler and clearer rules enabling authorities and those having to follow the rules to focus on preventing and eradicating diseases;
- clarifying responsibilities for farmers, veterinarians and others dealing with animals;
- promoting a greater use of new technologies for animal health such as surveillance of pathogens, electronic identification and registration of animals;
- allowing better early detection and control of animal diseases, including emerging diseases linked to climate change, to reduce the occurrence and effects of animal epidemics;
- ensuring more flexibility to adjust rules to local circumstances and to emerging issues such as climate and social changes; and
- ensuring a better legal basis for monitoring animal pathogens resistant to antimicrobial agents supplementing existing rules.

Several delegated and implementing acts have been and have still to be adopted by the European Commission to make the new rules on animal health fully applicable and to reach the objective of the Animal Health policy to raise the health status and improve the conditions of the animals in the EU, in particular, food-producing animals, whilst permitting intra-Community trade and imports of animals and animal products in accordance with the appropriate health standards and international obligations.

10.3 ANIMAL WELFARE

The general aim of the Animal Welfare policy is to ensure that animals avoid enduring avoidable pain or suffering and obliges the owner/keeper of animals to respect minimum welfare requirements. The EU zootechnical legislation aims at the promotion of free trade in breeding animals and their genetic material considering the sustainability of breeding programmes and the preservation of genetic resources. As it is stressed in the European Commission website, an important step in this domain has been the Council Directive 98/58/EC on the protection of animals kept for farming purposes, providing general rules for the protection of animals of all species kept for the production of food, wool, skin or fur or for other farming purposes, including fish, reptiles or amphibians. These rules are based on the European Convention for the Protection of Animals kept for farming purposes and reflect the so-called Five Freedoms from hunger and thirst, discomfort, pain, injury and disease fear and distress and to express normal behaviour.

When the EU Lisbon Treaty came into force in 2009, the recognition that animals are sentient beings was introduced (Article 13 of Title II). On 1 January 2012, conventional battery cages for laying hens were banned across the Union.

National governments may adopt more stringent rules, provided they are compatible with the provisions of the Treaty as Community legislation concerning the welfare conditions of farm animals lays down only minimum standards. As highlighted in the European Commission website, the enforcement of animal welfare legislation falls within the principle of subsidiarity, which means that:

- The EU Member States are responsible for:
  - day-to-day enforcement through their national legislation and control activities; and
  - transposition of directives into national legislation and the implementation of EU rules at the national level.
- The European Commission is responsible for:
  - providing appropriate information and, where necessary, training on EU legislative requirements;

- ensuring that EU legislation is properly implemented and enforced; and
- in extreme cases, taking action against EU countries that have failed to implement legal requirements.

These objectives are achieved through:

- inspections and controls undertaken by the Health and Food Audits and Analysis Directorate to check that competent authorities in EU countries apply EU legislation in an effective and uniform manner;
- the Standing Committee on the Food Chain and Animal Health that provides a platform for representatives of the Member States to discuss issues of public or animal health or animal welfare and when necessary approve urgent measures; and
- the EFSA collaboration for providing scientific opinions.

The legislation is designed to cover all stages of a farm animal's life whilst on the farm, during transport and at the time of killing.

10.4 PLANT HEALTH AND THE EUROPHYT NETWORK

The European Commission takes part actively in the setting of international phytosanitary and quality standards for plants and plant products. The EU legislation has also, over the years, provided for the harmonized protection of "green resources" (EU website, 2019). The reasons that have led to the review of the EU framework for plant health as set by Directive 2000/29/EC, which for almost two decades has served as a reference framework for the surveillance and the management of phytosanitary concerns throughout the EU, have been extensively reviewed by Montanari and Traon (2017). In this article, Montanari and Traon point out the importance for the modernization of the EU plant health of the EU enlargement and trade globalization together with some major institutional developments occurred following the adoption of Regulation (EC) 178/2002 and the EU adhesion to the International Plant Protection Convention (IPPC). In October 2016, the European Parliament and the Council adopted Regulation (EU) 2016/2031 on protective measures against plant pests ("Plant Health Law"). On 13 December 2016, the Regulation entered into force and is applicable since 14 December 2019. The new Regulation is taking care of pest categorization and prioritization. In line with this approach, plant pests are to be grouped into three main categories based on the risk they pose to plant health and/or quality, namely:

a) EU quarantine pests, including EU priority pests (Chapter II, Section II, Articles 4–31);
b) protected zone quarantine pests (Chapter II, Section III, Articles 32–35); and
c) regulated non-quarantine pests (Chapter III, Articles 36–39).

The new rules aim to modernize the plant health regime, enhancing more effective measures for the protection of the Union's territory and its plants. They also aim to ensure a safe trade, as well as to mitigate the impacts of climate change on the health of our crops and forests. Several delegated and implementing acts have to be adopted by the Commission to ensure the correct implementation of this legislation across the EU Member States. From 14 December 2019, all plants (including living parts of plants) need to be accompanied by a phytosanitary certificate to enter the EU unless they are listed in the Commission Implementing Regulation (EU) 2018/2019 as exempted from this general requirement. Currently, the list of plants exempted from the obligation to carry a phytosanitary certificate from 14 December 2019 includes the following fruits: pineapples, coconuts, durians, bananas and dates. The Plant Health Law increases the prevention against the introduction of new pests via imports from third countries. The Commission Implementing Regulation (EU) 2018/2019 establishes the list of high-risk plants, the introduction of which into the EU territory is to be provisionally prohibited from 14 December 2019 until a full risk assessment has been carried out. The rules concerning the procedure to be followed in order to carry out the risk assessment of high-risk plants are detailed in the Commission Implementing Regulation (EU) 2018/2019. Article 6(2) of Regulation (EU) 2016/2031 of the European Parliament and the Council on protective measures against pests of plants empowers the Commission to adopt delegated acts supplementing that Regulation by establishing a list of the priority pests. The delegated act was adopted by the Commission on 1 August 2019. A two-month scrutiny period then followed during which neither Council nor Parliament raised objections. Therefore, the Regulation has been published. The EC has recently published a list of 20 regulated quarantine pests qualified as priority pests, including *Xylella fastidiosa*, the Japanese beetle, the Asian long-horned beetle, citrus greening and citrus black spot, whose economic, environmental and social impact on EU's territory is the most severe. The Member States will have to launch information campaigns to the public; to do annual surveys; and to prepare contingency plans, simulation exercises and action plans for the eradication of these pests. The selection of pests is based on the assessment carried out by the Commission's Joint Research Centre and the European Food Safety Authority, which takes into account the probability of spreading, establishment and consequences of those pests for the Union.

Opinions of dedicated and public feedback provided via the Better Regulation Portal were also taken into account. The new methodology shows for example that the bacterium *X. fastidiosa*, the pest with the highest impact on agricultural crops, including fruit, could cause annual production losses of €5.5 billion, affecting 70% of the EU production value of older olive trees (over 30 years old) and 35% value of younger ones, in a scenario of the bacterium spreading across the entire EU. In addition to direct impacts on production, pests have significant indirect effects on a wide range of upstream or downstream economic sectors. For example, should the Asian long-horned beetle (*Anoplophora glabripennis*) spread across the entire EU, this could result in the direct loss of over 5% of the overall growing stock of several EU forestry tree species, such as alder, ash, beech, birch, elm, maple or plane trees. These trees are valued at €24 billion, and the economic impact on the upstream forestry sector could amount to €50 billion. The overall conclusion is that the new Regulations mainly addressed in this section have laid down the foundations for the strengthening of an EU phytosanitary legislation that is increasingly science-based and risk-oriented while remaining committed to free trade in plant and plant products (Montanari and Troan, 2017). It is also important in this context to mention the EUROPHYT network that is a web-based network and database. It connects plant health authorities of the EU Member States and Switzerland, the European Food Safety Authority and the Directorate-General for Health and Food Safety of the European Commission. As it is highlighted in the EC website, the main features of the EUROPHYT network are:

- **Notification of interceptions**: Plant health authorities of the EU Member States and Switzerland enter data about interceptions they have made of non-compliant consignments into EUROPHYT electronically, via a direct web-link.
- **A rapid alert system**: EUROPHYT immediately notifies the plant health authorities of the Member States and Switzerland of each interception. In the case of interceptions of imports from non-EU countries, the plant health authority of the exporting country also receives immediate notification in the form of an e-mail.

- **Database and information system**: All notifications are stored in a structured database. Members of the EUROPHYT network have full access to the data, making it possible to analyze trends and produce statistics.
- **Reports**: Standard weekly, monthly and annual reports are produced for different users.

10.5 ORGANIC FOOD

Organic food is food produced by methods that comply with the standards of organic farming. As stated in the European Commission website, organic farming is an agricultural method that aims to produce food using natural substances and processes. This means that organic farming tends to have a limited environmental impact as it encourages the:

- responsible use of energy and natural resources;
- maintenance of biodiversity;
- preservation of regional ecological balances;
- enhancement of soil fertility; and
- maintenance of water quality.

Additionally, organic farming rules encourage a high standard of animal welfare and require farmers to meet the specific behavioural needs of animals. EU Regulations on organic farming are designed to provide a clear structure for the production of organic goods across the whole of the EU. This is to satisfy consumer demand for trustworthy organic products whilst providing a fair marketplace for producers, distributors and marketers. Regulation (EU) 2018/848 of 30 May 2018 on organic production and labelling of organic products was published on 14 June 2018 and applies from 1 January 2021. It repeals Regulation (EU) 834/2007 and introduces several new features (Phytocontrol, 2018):

- hydroponic production, which consists in placing the roots of certain plants in a nutrient solution, is prohibited;
- the production of processed organic food excludes food containing or consisting of manufactured nanomaterials;
- some products have been included in the scope of the Regulation to the extent that they can be produced using natural production techniques (see Annex I of this Regulation);
- a group certification system for small farmers and operators has been set up to reduce inspection and certification costs and administrative constraints, strengthen local networks, contribute

to the development of better market outlets and ensure a level playing field with third-country operators;
- a farm can produce organic, in-conversion and non-organic products under certain conditions;
- operators and groups of operators shall be subject to a compliance check at least once a year. In the case where no non-compliance has been found for three consecutive years, the interval between two on-site physical inspections shall be two years; and
- by 31 December 2024 at the latest, the Commission will submit a report to the European Parliament and the Council on the presence of products and substances not authorized for use in organic production and on the evaluation of existing national rules. This report may be accompanied by a legislative proposal for further harmonization. At present, the Member States which have national rules prohibiting products with an unauthorized substance content above a certain level from being marketed as organic may continue to apply those rules, provided that they do not prohibit, restrict or prevent the placing on the market as organic products of products obtained in the other Member States, where those products have been produced in accordance with this Regulation. Member States which make use of these rules must inform the European Commission. In the event of contamination by an unauthorized substance, if the inspection body establishes that it has been used voluntarily or that precautionary measures have not been applied, the product concerned shall not be marketed as an organic product or in conversion or used in organic production.

Regulation (EU) 2018/484 also provides for complex rules for dealing with traces of agrochemicals in organic products (Schmidt, 2019). Organic farmers are now to draw up and maintain "precautionary measures" against pollution from conventional agriculture. This Regulation introduces three legal consequences for each case where a pesticide trace is reported in an organic food product, however low this might be: (a) an official investigation; (b) a provisional marketing stop; or (c) permanent organic decertification.

10.6 COUNTRY OF ORIGIN OR PLACE OF PROVENANCE

For an international perspective on the country of origin labelling rules for prepacked foods and ingredients, see the specific reference

(Carbonelle at al, 2016). The EU has established mandatory origin information provisions for certain products under product-specific rules, e.g. beef and beef products, fishery products (Reg. (EU) 1379/2013), fruit and vegetables (Reg. (EU) 1308/2013), honey (Directive 2014/63/EU) and olive oil (Reg. (EU) 29/2012). Foods which do not have product-specific provisions regarding the country of origin or place of provenance indication will be subject to the EU's non-preferential rules of origin, i.e. the customs legislation. The Article 26 of Regulation (EU) 1169/2011 shall apply without prejudice to labelling requirements provided for in specific Union provisions, in particular Council Regulation 509/2006 of 20 March 2006 on agricultural products and foodstuffs as traditional specialties guaranteed and Council Regulation (EC) 510/2006 of 20 March 2006 on the protection of geographical indications and designations of origin for agricultural products and foodstuffs. Indication of the country of origin or place of provenance becomes mandatory where failure to indicate this may mislead the consumer as to the true country of origin or place of provenance of the food, in particular if the information accompanying the food or the label as a whole would otherwise imply that the food has a different country of origin or place of provenance. According to the EC website, the Commission Implementing Regulation (EU) 1337/2013 of 13 December 2013 lays down rules for the application of Regulation (EU) 1169/ 2011 of the European Parliament and the Council as regards the indication of the country of origin or place of provenance for fresh, chilled and frozen meat of swine, sheep, goats and poultry (Tables 10.1–10.4). This Regulation entered into force on 1 April 2015.

Additional voluntary indications (e.g. flags) remain permitted as long as they are not misleading.

On 12 May 2016 the European Parliament adopted the resolution on the mandatory indication of the country of origin or place of provenance for certain foods. In that resolution the European Parliament (see Annex 1):

- called on the Commission to implement the mandatory indication of the country of origin or place of provenance for all kinds of drinking milk, dairy products and meat products, and to consider

Table 10.1 Indication of the Country of Origin or of the Place of Provenance for Poultry Meat

Country of the last period of rearing of at least one month

If slaughtered younger than one month: country of the all rearing period

152

Table 10.2 Indication of the Country of Origin or of the Place of Provenance for Swine Meat

Less than six months of age	Country of the last four months of rearing
Less than six months of age with more than 80 kg live body weight	Country in which the swine has been reared since a live body weight of 30 kg until slaughtering
Less than six months of age with less than 80 kg live body weight	Country in which the whole rearing has been carried out since birth
More than 6 months of age	Country in which rearing has been carried out during the last six months of life

Table 10.3 Indication of the Country of Origin or of the Place of Provenance for Sheep and Goat Meats

More than six months of age	Country where the last six months of rearing have taken place
Less than six months of age	Country where the all rearing period since birth has taken place

extending the mandatory indication of the country of origin or place of provenance to other single-ingredient foods or those with one main ingredient, by making legislative proposals in these areas;

- urged the Commission to submit legislative proposals, making the indication of the origin of meat in processed foods mandatory in order to ensure greater transparency throughout the food chain and to better inform European consumers. Mandatory labelling requirements should take into account the principle of proportionality and the administrative burden for food business operators and enforcement authorities;
- considered that the aim of mandatory food origin labelling is to restore consumer confidence in food products;
- called on the Commission to make a proposal to this end while taking into account the transparency of the information and its legibility for consumers, the economic viability of European businesses and the purchasing power of consumers;
- called on the Commission to support labelling schemes relating to animal welfare during cultivation, transport and slaughter;

Table 10.4 Derogations for Minced Meats and Trimmings

For meats coming from third countries

"Reared in non-EU" and slaughtered in; the name of the country in which slaughtering has taken place"

For minced meats and trimmings

(a) "EU Origin" if minced meats or trimmings are produced only with meats obtained from animals born, reared or slaughtered in one or more Member States

(b) "Reared and slaughtered in: EU" if minced meats and trimmings have been produced exclusively with meats from animals reared and slaughtered in one or more Member States

(c) "Reared and slaugthered in: non-EU" if minced meats and trimmings have been produced exclusively with meats imported in the Union

(d) "Reared in: non-EU" and "Slaugthered in EU" if minced meats or trimmings have been produced exclusively with meats obtained from animals imported in the Union as animals for slaughtering and slaughtered in one or more Member States

(e) "Reared and slaughtered in: EU and non-EU"

In case the minced meats and trimmings have been produced with:

(i) meats obtained from animals reared and slaughtered in one or more Member States and from meats imported in the Union

(ii) meats obtained from animals imported in the Union and slaughtered in one or more Member States

(e) "Reared and slaughtered in: EU and non-EU"

In case the minced meats and trimmings have been produced with:

(i) meats obtained from animals reared and slaughtered in one or more Member States and from meats imported in the Union

(ii) meats obtained from animals imported in the Union and slaughtered in one or more Member States

- deplores the fact that the Commission had still not made any move to include eggs and egg products in the list of foods for which the indication of the country of origin or place of provenance is mandatory; and
- believed that country of origin labelling for drinking milk, lightly processed dairy products (such as cheese and cream) and lightly

processed meat products (such as bacon and sausages) would have significantly reduced associated costs and that this labelling should be explored as a priority.

Article 26(3) of Regulation (EU) 1169/2011 requires that where the origin of a food is given and is different from one of its primary ingredients, the origin of the primary ingredient shall be given or at least indicated as being different from the origin of the food. The Commission Implementing Regulation (EU) 2018/775 of 28 May 2018 lays down rules for the application of Article 26(3) of Regulation (EU) 1169/2011 on the provision of food information to consumers, as regards the rules for indicating the country of origin or place of provenance of the primary ingredient of a food. A "primary ingredient" is an ingredient (or ingredients) of a food that represents more than 50% of the food or that is usually associated with the name of the food by the consumer and for which in most cases a quantitative indication (QUID%) is required. This Regulation does not apply to geographical indications protected under Regulation (EU) 1151/2012, Regulation (EU) 1308/2013, Regulation (EC) 110/2008 or Regulation (EU) 251/2014 or protected pursuant to international agreements, nor registered trademarks where the latter constitute an indication of origin, pending the adoption of specific rules concerning the application of Article 26(3) to such indications. The country of origin or the place of provenance of a primary ingredient which is not the same as the given country of origin or the given place of provenance of the food shall be given:

(a) with reference to one of the following geographical areas:
   (i) "EU", "non-EU" or "EU and non-EU";
   (ii) region, or any other geographical area either within several Member States or within third countries, if defined as such under public international law or well understood by normally informed average consumers;
   (iii) FAO Fishing area, or sea or freshwater body if defined as such under international law or well understood by normally informed average consumers;
   (iv) Member State(s) or third country(ies);
   (v) region, or any other geographical area within a Member State or within a third country, which is well understood by normally informed average consumers; or
   (vi) the country of origin or place of provenance in accordance with specific Union provisions applicable for the primary ingredient(s) as such "(name of the primary ingredient) do/

does not originate from (the country of origin or the place of provenance of the food)" or any similar wording likely to have the same meaning for the consumer.

The information provided pursuant to Article 2 shall be provided in a font size which is not smaller than the minimum font size as required in accordance with Article 13(2) of Regulation (EU) 1169/2011.

Where the country of origin or place of provenance of a food is given with words, the information provided shall appear in the same field of vision as the indication of the country of origin or place of provenance of the food and by using a font size which has an x-height of at least 75% of the x-height of the indication of the country of origin or place of provenance of the food.

Where the country of origin or place of provenance of a food is given by means of non-scriptural form, the information provided shall appear in the same field of vision as the indication of the country of origin or place of provenance of the food. This Regulation applies from 1 April 2020. On 30 January 2020, the Commission adopted a "notice" on the application of the provisions of Article 26(3) of Regulation (EU) 1169/2011 with regard to the origin indication of the primary ingredient of a food. It aims at assisting all players in the food chain as well as the competent national authorities to better understand and correctly apply the provisions of Regulation (EU) 1169/2011 related to the origin indication of the primary ingredient (European Commission website).

Regulation (EC) 1760/2000, replacing the Council Regulation 820/1997, was adopted by the Parliament and the Council on 17 July 2000 to establish a system for the identification and registration of bovine animals and for the labelling of beef and beef products.

11

The European Food Safety Authority

11.1 INTRODUCTION

EFSA was established by Regulation (EC) 178/2002 as the European Scientific Agency in charge of risk assessment and risk communication for all the food/feed chain as well as for animal and plant health. The European Institutions have recently adopted a new Regulation to increase the transparency and sustainability of the EU risk assessment. This is Regulation (EU) 2019/1381 adopted on 20 June 2019 by the European Parliament and the Council on the transparency and sustainability of the EU risk assessment in the food chain and amending Regulations (EC) 178/2002, (EC) 1829/2003, (EC) 1831/2003, (EC) 2065/2003, (EC) 1935/2004, (EC) 1331/2008, (EC) 1107/2009, (EU) 2015/2283 and Directive 2001/18/EC. Among others, this new Regulation has amended some articles of Regulation 178/2002, including some dealing with structural or procedural characteristics of EFSA (see Chapter 12). In general, this Regulation applies from 18 months since its entry into force, whereas specific sections (i.e. Articles 1(2) and 1(3)) will apply later. For the ongoing changes in the EFSA status described in this chapter depending on this new Regulation see Chapter 12.

11.2 MAIN EFSA TASKS

The main EFSA's tasks aim to:

i) provide the Community institutions (i.e. European Commission, Parliament and the Member States) with the best possible advice on risk assessment;

ii) promote and coordinate the development of risk assessment methodologies in different sectors;

iii) provide scientific and technical support to the European Commission under normal conditions and also in case of crisis;

iv) identify and characterize emerging risks with a view to facilitating their prevention; and

v) ensure that the public and interested parties receive rapid, reliable, objective and comprehensible information in the fields within EFSA's mission.

Clearly EFSA's activity is strictly linked with science and focused on risk assessment and scientific advice on food and feed-related risks, dietary issues, animal health and welfare and plant health and related risk communication.

On the other hand, EFSA does not work on:

(i) food safety policies standards;

(ii) risk management and

(iii) enforcement/control.

11.3 EFSA INDEPENDENCE POLICY

In October 2016, the management board set up a working group on the review of EFSA's independence policy. The working group met between November 2016 and January 2017 and reviewed the current EFSA Policy on Independence primarily in the areas of:

- definition of conflict of interest;
- financial or economic interests;
- risk-based approach for competing interest management;
- cooperation with EFSA's partners;
- cooling-off periods;
- research funding and other scientific activities;
- transparency; and
- enforcement.

The current policy to ensure EFSA's independence, adopted in June 2017, is based on the adoption and implementation of conflict of interest management rules applicable to experts or to staff members.

In 2018 several documents were published:

- publication of a register of activities undertaken by former members of the management board for two years after their term of office ended;
- publication of the list of "Public Institutions";
- publication of EFSA middle management DOIs; and
- publication of the first annual report on independence activities.

11.4 EFSA ORGANIZATIONAL FRAMEWORK

Several approaches have been adopted when establishing EFSA in order to ensure its independence from all interested parties in its opinions, including risk managers.

11.4.1 People Working for EFSA

With reference to EFSA's organizational setting, several approaches have been adopted when establishing EFSA in order to ensure its independence from all interested parties in its opinions, including risk managers. Scientists providing opinions for EFSA as member of the Scientific Committee or the Scientific Panels are not EFSA's employees, whereas EFSA's staff (about 450 units) supports and facilitates the work of these experts within the EFSA organizational framework. EFSA experts are coming from the academia and other scientific institutions within and outside the EU (e.g. third countries) and are appointed in each EFSA's Scientific Panel or the Scientific Committee for a term of three years, following a selection process, mainly based on proven scientific excellence within needed disciplines, among all the responders with their CVs to a specific EFSA call for the manifestation of interest published every three years. About 15 qualified scientific experts are assembled in each one of the 10 panels with sectoral competences (see section 5.4.2), and the Scientific Committee, competent for scientific coordination and multisectoral tasks, is formed by all the chairs of the ten panels and six additional *ad hoc* selected members. Independent EFSA experts in the scientific Committee and panels develop risk assessment methodologies in the areas of competence and carry out scientific assessments applying them to specific cases. Appointed experts may remain in a specific office for only three consecutive terms, in any case following three renewed *ad hoc* application and procedures. The experts appointed in specific panels/committee and working groups must be able to operate with independence and be free from any interest in contrast with EFSA policies. *Ad hoc* policy for avoiding conflicts of interest of the experts appointed in specific panels/committee and working groups have been adopted.

11.4.2 EFSA Administrative and Scientific Bodies

EFSA is governed by a management board with 15 members. The board sets EFSA's budget and approves its annual work programme. EFSA's executive director is responsible for operational and staffing matters. He also draws up the annual work program, together with the Commission, the European Parliament and the EU Member States. The Advisory Forum advises the executive director. EFSA is structured into a Scientific Committee and 10 Scientific Panels. The Scientific Committee, in addition to being responsible for the general coordination to ensure consistency in the scientific opinions prepared by the Scientific Panels, has also the task of supporting the work of EFSA on scientific matters of a horizontal nature and providing strategic advice to EFSA's executive director (Silano et al, 2012).

As described in the EFSA website, the mandates of the ten Scientific Panels are as follows:

- AHAW – the Panel on Animal Health and Welfare provides scientific advice on all aspects of animal diseases and animal welfare. Its work chiefly concerns food-producing animals, including fish.
- BIOHAZ – the Panel on Biological Hazards provides scientific advice on biological hazards in relation to food safety and foodborne diseases. This covers animal diseases transmissible to humans, transmissible spongiform encephalopathies, food microbiology, food hygiene and associated waste management issues.
- CEP – the Panel on Food Contact Materials, Enzymes and Processing Aids evaluates the safety of chemical substances added to food or used in food packaging and related processes. The Panel's work mainly concerns substances and processes evaluated by EFSA before their use can be authorized in the EU.
- CONTAM – the Panel on Contaminants in the Food Chain provides scientific advice on contaminants in the food chain and

undesirable substances such as natural toxicants, mycotoxins and residues of unauthorized substances.

- FAF – the Panel on Food Additives and Flavourings evaluates the safety of chemical substances added to foods as additives or flavourings and consumer exposure to them. The Panel's work mainly concerns substances evaluated by EFSA before their use can be authorized in the EU.
- FEEDAP – the Panel on Additives and Products or Substances used in Animal Feed provides scientific advice on the safety and/ or efficacy of additives and products or substances used in animal feed. The Panel evaluates their safety and/or efficacy for the target species, the user, the consumer of products of animal origin and the environment. It also looks at the efficacy of biological and chemical products/substances intended for deliberate use in animal feed.
- GMO – the Panel on Genetically Modified Organisms provides scientific advice on food and feed safety, environmental risk assessment and molecular characterization/plant science. Its work chiefly concerns genetically modified plants, micro-organisms and animals.
- NDA – the Panel on Nutrition, Novel Foods and Food Allergens deals with questions related to human nutrition, novel foods, nutrient sources, foods for special groups such as infant formulae, health claims on food products, dietary reference values and food allergies.
- PLH – the Panel on Plant Health provides independent scientific advice on the risk posed by plant pests which can cause harm to plants, plant products or biodiversity in the EU. The Panel reviews and assesses those risks with regard to the safety and security of the food chain.
- PPR – the Panel on Plant Protection Products and Their Residues provides scientific advice on the risk assessment of pesticides for operators, workers, consumers and the environment. The Panel develops and reviews guidance documents on the risk assessment of pesticides. This work supports the evaluation of active substances used in pesticides, which is carried out by Rapporteur Member States and peer-reviewed by EFSA staff.

If specific knowledges are needed for dealing with a specific subject, the Scientific Committee or a Panel may set up a working group. These working groups may include both EFSA scientists and external experts selected as needed by making use of EFSA's expert database, including thousands of experts.

EFSA also collaborates closely with other EU agencies active in the field of health and safety issues relating to humans, animals and the environment:

- European Medicines Agency (EMA);
- European Chemicals Agency (ECHA);
- European Centre for Disease Prevention and Control (ECDC); and
- European Environmental Agency (EEA).

EFSA's overall structure has been slightly variable over time, but, in addition to the ten Panels, generally structured in one or two departments, has also included other departments such as "Communication" and "Administration". EFSA became operational in 2003. It grew rapidly (the budget increased from €10.3 million in 2003 to a stable budget of around €79 million, and its staff increased from 72 in 2003 to 445 in 2014, 74% being allocated to operational activities and 26% to support activities). Its workload increased rapidly over time (more applications for authorizations) due to a high number of new legal texts on authorizations adopted since 2003. A sharp increase (74.5%) in the number of requests for scientific opinions sent to EFSA was observed between 2006 and 2011.

11.4.3 EFSA Application Helpdesk

The "application helpdesk" is the EFSA's front office and support desk to assist the submission and the monitoring of the applications on regulated products, substances and processes that require a scientific risk assessment by EFSA, before they can be authorized by risk managers, and the substantiation of claims submitted for authorization in the EU.

The application helpdesk is the first point of contact with EFSA for applicants, the Member States and any stakeholder regarding applications on regulated products and provides information on:

- the legal framework for each application type (e.g. food additives);
- procedures for submission of technical dossiers;
- EFSA guidance documents per scientific area;
- administrative and scientific data requirements;
- information on the processing and status of the applications;
- frequently asked questions of the scientific area; and

- the Register of Questions database that collect all the information about all EFSA's work, including documents and current status of applications.

The application helpdesk can be contacted by sending an enquiry via the dedicated web form. The helpdesk will either answer the question directly liaising as needed with the scientific unit or guide the applicant to an appropriate contact point.

The helpdesk may support requestors by guiding them through the relevant legislative framework, the updates on the processing and status of the application. In partnership with EFSA's scientific units, it may also organize technical meetings with stakeholder groups with the aim to increase the understanding of EFSA guidance documents and to improve the quality of the submitted applications.

11.5 THE EFSA STRATEGY, ITS UPDATING FOR THE PERIOD 2021–2027 AND TRAINING ACTIVITY

Currently, EFSA is operating under the EFSA 2020 strategy, an evolutive approach to current challenges that does not stand in isolation but has made extensive use of existing planning and past experience. Five over-arching "strategic objectives" have characterized the EFSA action during the last years (Silano, 2014b):

- prioritize public and stakeholders engagement in the process of scientific assessment.
  Four operational objectives have been adopted to ensure the achievement of this strategic objective:
    - promote an enhanced dialogue with stakeholders on man-dates in collaboration with risk managers;
    - make documentation on information gathering and the evalu-ation process available;
    - foster engagement throughout the development of scientific assessment; and
    - ensure clarity and accessibility/usability in the communica-tion of findings.
- widen evidence base and optimize access to data.
  Three operational objectives have been adopted to ensure the achievement of this strategic objective:
    - to adopt an open-data approach;

- – to improve data interoperability to facilitate data exchange;
  and
- – to migrate towards structured scientific data;
- • build the EU's scientific assessment capacity and knowledge
  community.
  Three operational objectives have been adopted to ensure the
  achievement of this strategic objective:
  - – strengthening capacity building and capacity sharing with
    the Member States in collaboration with the European
    Commission Directorate–General for Research and innovs-
    tion and its Joint Research Centre, EU Agencies and interna-
    tional organisations;
  - – fostering the growth of the EU risk assessment community in
    collaboration with international organizations; and
  - – reviewing and further developing EFSA's scientific assess-
    ment models.
- • prepare for future risk assessment challenges.
  Three operational objectives have been adopted to ensure the
  achievement of this strategic objective:
  - – to strengthen EFSA's resilience and ability to anticipate and
    respond effectively to food safety risks in cooperation with
    EU and international partners;
  - – to develop and implement harmonized methodologies and
    guidance documents for risk assessment across the EU and
    internationally; and
  - – to become a hub in methodologies, tools and guidance docu-
    ments for risk assessment;
- • create an environment that reflects EFSA's.
  Two operational objectives have been adopted to ensure the
  achievement of this strategic objective.
  - – people: build a culture that puts EFSA values into practice;
    and
  - – organization and processes: develop an environment focused
    on improving organizational performance and capabilities.

It is also important to stress that the European Food Safety Authority
(EFSA) has recognized a need for training its Networks, Panel and
Scientific Committee members and scientific staff to facilitate their under-
standing, uptake and use of the best risk assessment practices developed

by EFSA (see also Chapter 13). In addition there is a need to strengthen the cross-fertilization of scientific discussion and support harmonization between the different scientific domains in EFSA. In cooperation with qualified scientific institutions, EFSA offers a series of specialized short training courses on specific aspects of food safety risk assessment for members of EFSA's Scientific Committee/Panels, their working groups, members of the EFSA Networks and EFSA staff.

As seating is limited for the in-house training courses, some aspects of most topics are also trained through *ad hoc* webinars. Detailed information on each course is made available via the website. In addition to a number of training courses, EFSA has also carried out study visitors and seconded national experts and innovation prize programmes. The development of EFSA activities in different areas has been reviewed at different times by Silano and Silano (2008); Deluyker and Silano (2012); and Silano et al. (2014b).

For further information on the EFSA Strategy 2020, it is possible to consult the EFSA Publication "EFSA Strategy 2020 – Trusted Science for Safe Food-Protecting Consumers' Health with Independent Scientific Advice on the Food Chain" that also includes in the Annex the implementation plans for all the five strategic objectives. An update of the EFSA Strategy for the years 2021–2027 is currently being discussed by the Scientific Committee based on the feedback from the EFSA management board and the results of the survey on horizon scanning. In this context, it may be also important to consider the current tendency in the EU to develop partnerships to promote a common understanding of risk assessment and to be ready to better face the future challenges. In autumn 2019, the Commission services asked potential partners to further elaborate proposals for European Partnership under Horizon Europe, among others, with specific reference to the Assessment of Risk from Chemicals (PARC).

11.6 RISK ASSESSMENT: APPROACHES, TOOLS AND RESOURCES

Risk assessment is the main task of EFSA and consists of:

- hazard identification and hazard characterization (Figure 11.1);
- exposure assessment; and
- risk characterization (Figure 11.2).

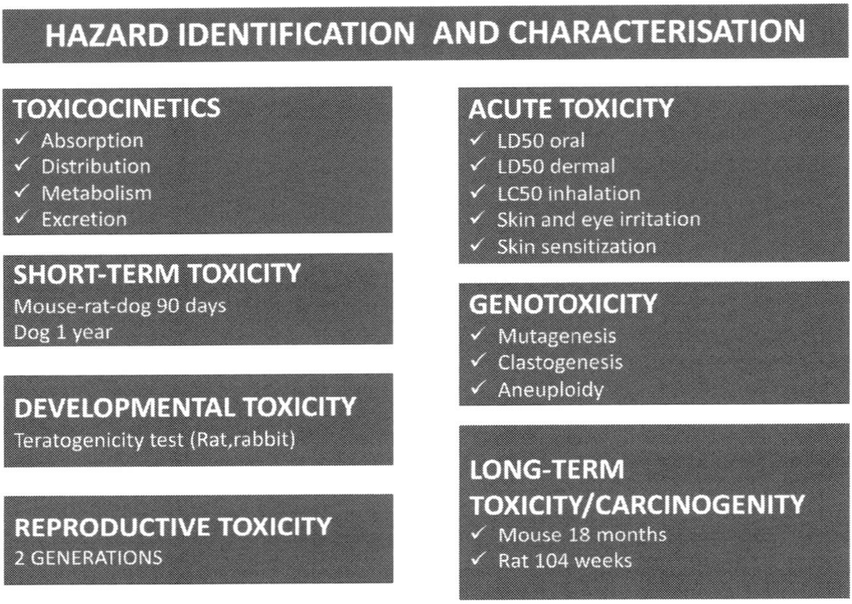

HAZARD IDENTIFICATION AND CHARACTERISATION

TOXICOCINETICS
- ✓ Absorption
- ✓ Distribution
- ✓ Metabolism
- ✓ Excretion

SHORT-TERM TOXICITY
Mouse-rat-dog 90 days
Dog 1 year

DEVELOPMENTAL TOXICITY
Teratogenicity test (Rat,rabbit)

REPRODUCTIVE TOXICITY
2 GENERATIONS

ACUTE TOXICITY
- ✓ LD50 oral
- ✓ LD50 dermal
- ✓ LC50 inhalation
- ✓ Skin and eye irritation
- ✓ Skin sensitization

GENOTOXICITY
- ✓ Mutagenesis
- ✓ Clastogenesis
- ✓ Aneuploidy

LONG-TERM TOXICITY/CARCINOGENITY
- ✓ Mouse 18 months
- ✓ Rat 104 weeks

Figure 11.1   Hazard identification and characterization.

In risk assessment, the toxicological evaluation is useful:

- to establish the intrinsic chemical or biological hazard in order to evaluate which exposure levels might be considered safe;
- to establish the exposure level likely to occur in relation to the use or occurrence in the diet; and
- to compare the daily exposure with the safe daily exposure level to check whether the level of risk is acceptable in the light of all available data.

The classical approach to assess the risk consists of the calculation of the ADI that represents the amount of a substance, generally expressed in mg/kg body weight, that can be ingested daily for the all lifetime, without any significant risk. It can be calculated by dividing the NOAEL by the uncertainty (or safety) factor.

ADI = acceptable daily intake (mg/kg b.w. and day)
NOAEL = no observed adverse effect level (mg/kg b.w. and day)
UF = uncertainty factor (100 or other values)

166

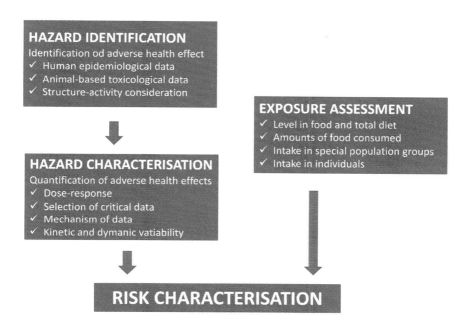

Figure 11.2   Risk characterization.

Many other different methodologies may be used for risk assessment depending on the nature of the risk to be assessed and of the available data. For instance, see the EFSA scientific publications on:

- BMD approach;
- harmonized approach for the assessment of substances which are genotoxic and carcinogenic;
- margin of exposure approach (MoE);
- threshold of toxicological concern (TTC); and
- reference points for action (RPAs).

To carry out risk assessment throughout the whole food/feed chain, EFSA's Scientific Committee and Panels have adopted many innovative risk assessment guidance documents and have applied them to thousands of cases producing so far a very large number of risk assessment methodologies and opinions.

EFSA risk assessment opinions and other documents adopted by the Scientific Committee or Panels in response to requests from the European Commission, the Parliament or the Member States or decided through self-tasking are adopted by consensus (minority opinions, if any, are always noted, with their motivations, in the text of the opinion). Each risk assessment opinion is published immediately after adoption in the *EFSA Journal*, in a standard format making possible the full understanding of the available data of the underlying scientific motivations and reasoning. A considerable number of assessment tools and resources have been developed and commissioned by EFSA's scientists, statisticians and technical staff to allow the rapid processing of information, support consistent analysis and reporting of findings and allow for predictive modelling as scientific assessments often require the analysis of large amounts of data. Sophisticated modelling techniques are established in some cases as alternative approaches when scientific data are lacking or non-existent.

Several repositories, platforms and assessment calculation tools are available for use by those involved in food and feed safety assessments (Table 11.1).

The European Food Risk Assessment Fellowship (EU-FORA) programme was born in 2016 as a part of EFSA efforts to build and strengthen Europe's risk assessment capacity and to promote harmonization of methodologies across organizations and countries.

11.7 SPECIFIC FEATURES OF EFSA RISK ASSESSMENT: CROSS-CUTTING GUIDANCE DOCUMENTS

EFSA's approach to risk assessment consists of applying a relevant formally adopted methodological guidance following the publication in its scientific journal. The adoption of risk assessment guidance documents takes place often through a public consultation process, and any comment not accepted is subject to a specific motivation by EFSA.

Scientific guidance documents are cross-cutting if they deal with matters of a horizontal nature in fields that affect scientific areas covered by more than one Panel. The main objective of developing cross-cutting guidance documents is to harmonize scientific risk assessment throughout EFSA as much as possible. The responsibility for elaborating and adopting cross-cutting guidance documents is with the Scientific Committee, whereas that for the guidance documents concerning the activity of only one Panel is with the specific Panel. On 16 April 2015, to increase the transparency and harmonization of EFSA opinions adopted by the Panels, EFSA asked its SC for advice on how cross-cutting guidance documents should be used across EFSA, reviewed and kept up-to-date.

Table 11.1   Repositories, Platforms and Assessment

1. Repositories and platforms

- **Knowledge Junction | Zenodo**

  The Knowledge Junction is a curated, open repository for the exchange of evidence and supporting materials used in food and feed safety risk assessment. It includes databases, models, reports and literature reviews. Anyone can upload resources on Knowledge Junction following a few simple steps. Its aim is to improve transparency, reproducibility and evidence re-use.

- **EFSA R4EU**

  The EFSA R4EU platform hosts a suite of different modules with different functionalities for modelling commonly used in assessments by EFSA. Existing models include benchmark dose, multi-drug resistance analysis, risk assessment using Monte Carlo, risk-based surveillance systems, sample size calculator, exploratory analysis for spatio-temporal epidemiology and automatic abstract and full-text screening using machine learning.

- **EFSA: API Developer Portal [beta]**

  The EFSA:API "developer portal" uses the application programming interface (API) technology to make EFSA's IT resources more accessible to software developers, allowing them to design creative new apps and tools for use by the food safety assessment community.

- **EFSA vector-borne disease map journals**

  The EFSA vector-borne diseases map journal provides detailed maps and information for 36 vector-borne diseases. This includes the disease agent, transmission, geographic distribution, potential vectors involved, impact on animal health and welfare and available prevention and control measures.

2. Assessment calculation tools

- **Feed Additive Consumer Exposure (FACE)**

  The FACE calculator estimates chronic and acute dietary exposure to residues of feed additives and their metabolites present in the food of animal origin.

- **Feed Additives Maximum Safe Concentration in Feed for Target Species calculator (FACTS)**

  The FACTS tool estimates the maximum safe concentration of feed additives in feed for different animal categories and species.

- **Food Enzyme Intake Model (FEIM)**

  The FEIM tool estimates chronic dietary exposure to food enzymes used in food processes. It uses summary statistics on food consumption data collected from Member States.

(Continued)

Table 11.1 (Continued) Repositories, Platforms and Assessment

- **Food Additives Intake Model 2.0 (FAIM)**
  The FAIM tool estimates chronic dietary exposure to food additives, including new food additives or authorized food additives for which a new use is proposed. Results are provided for different population groups (e.g. infants, toddlers, adults, etc.) and for different countries.

- **PRIMo and other pesticide evaluation tools**
  Several pesticide-related calculation tools and models support assessors, managers and applicants on pesticide evaluations, including pesticide residues intake model (PRIMo), pesticide fate models, non-dietary (operator, worker, resident and bystander) exposure to pesticides, dermal absorption and risks for bees.

- **Rapid Assessment of Contaminant Exposure (RACE)**
  The RACE tool provides estimates of different population groups' acute and chronic exposure to chemical contaminants from single foods and compares the result to the health-based guidance value or other relevant toxicological reference points.

- **Risk assessment for infectious diseases in animals (MINTRISK)**
  The Method for INTegrated RISK (MINTRISK) assessment for infectious diseases in animals allows the risk assessment of vector-borne diseases of livestock and pets. Developed by Wageningen BioVeterinary Research and Wageningen Economic Research, it has specific functions for assessment in EFSA's remit.

Source: EFSA website.

The EFSA SC recommended that each guidance document identified its target audience and addressed the level of obligation to follow. The EFSA SC also recommended that cross-cutting guidance documents should be screened for their up-to-date scientific relevance every three years. The EFSA document to present the lifecycle of cross-cutting guidance documents deals with:

- development;
- implementation (monitoring);
- review; and
- revision.

The cross-cutting methodologies adopted by EFSA so far are currently visible on the EFSA website, and a considerable number of cross-cutting documents have to be included in the ongoing post-adoption monitoring programme. Main steps include:

- record of dissemination channels used;
- record of capacity building:
  - feedback on the use of guidance documents – panels/units;
  - working groups tracking the use of guidance documents in EFSA scientific assessments; and
  - feedback on the use of guidance documents – cross-cutting working groups.

For tracking the use of guidance documents in EFSA scientific assessments, a valid output is represented by:

- the list of scientific opinions in which given guidance is cited (including information on the Panel/Unit publishing this scientific opinion); and
- the number of scientific opinions in which a given Guidance Document is cited/number of scientific opinions published in the period of interest (overall results per Panel and Unit).

11.8 EMERGING RISKS

Emerging risk identification is a very demanding task for EFSA that was assigned in its founding Regulation (Altieri et al., 2011; Robinson et al., 2012). The importance of the activities on emerging risks derives from the possibility to anticipate and even, in some case, to prevent future food/feed safety challenges. The ability to discover emerging risks is based on available data/knowledge and adequate emerging risk assessment methods.

As defined by EFSA, an "emerging issue" is one that has very recently been identified and deserves further investigation, and for which the information collected is still too limited to be able to assess whether it meets all the requirements of an emerging risk. Thus, "emerging issues" may be identified at the beginning of the emerging risk identification process as subjects that deserve further investigation and additional data collection and analysis (EFSA, 2015). The identification of emerging risks or emerging issues implies a complex process requiring broad expertise and close cooperation of EFSA with the Member States, food business operators and international and European agencies. Dedicated networks provide the structures needed to exchange experience, methods and data and to assess emerging issues. The EFSA network of knowledge includes the:

- Emerging Risks Exchange Network;
- stakeholders risks;

- EFSA Scientific Units and Scientific Panels; and
- Scientific Committee and its working Groups.

An international collaboration has also been established with FAO and WHO and with the participation in the International Liaison Group on Chemical and Microbiological Food Safety.

One useful approach to identify emerging risks is to closely follow changes in:

- pests/pathogens, vectors and or hosts (plants/animals) and their interactions;
- agriculture and forestry practices/husbandry production;
- trade, food consumption and land use; and
- climate changes.

EFSA's activity on emerging risks has resulted in:

- development of ERI procedure and methodologies;
- identification of a number of relevant issues;
- developing expertise and networking;
- sharing information without creating unnecessary scares; and
- focused monitoring and follow-up activities.

The main objectives of EFSA's activities on emerging risks are (i) to identify emerging risks in the areas within the remit of EFSA; and (ii) to develop and improve emerging risk identification methodologies and approaches. A number of different projects on emerging risks have been carried out by EFSA, so far. A very recent project (REACH 3) based on the application of the procedure for screening chemicals registered under the ECHA Regulation has resulted in the identification of 212 "potential emerging risks/emerging chemical issues" which were identified as chemicals being produced in large amounts and in conditions making possible the release to the environment (Oltmanns et al., 2019). These chemical substances were qualified as non-degradable, bio-accumulable in different segments of the food chain and characterized by known toxicity. However, an in-depth evaluation, carried out on a sample of 10 out of the 212 substances, to establish the possible occurrence of these substances in the food chain as contaminants did not allow to conclude on an "emerging risks" status for these substances. Similar limitations experienced during evaluations of the selected 10 "potential emerging risks" likely apply to most of the above-mentioned 212 substances. Therefore, the current main question concerns which ones of these 212 substances currently occur in the food chain as a contaminant and their levels. The future work programme consists of:

- assessment of occurrence in food/feed and/or environmental media through suspect and non-target screening approaches; and
- use of different and highly sensitive analytical methods (e.g. LC–MS and GC–MS) in different (raw) food and feed items;

to identify substances in the priority list for more exhaustive monitoring.

Another emerging risk project (AQUARIUS) has been carried out by EFSA between December 2015 and March 2019 on the applicability of global food chain analysis for identification of vulnerabilities and drivers of change. This project has three main blocks:

- Block 1: Mapping of the salmon supply chain;
- Block 2: Vulnerabilities, emerging hazards (animal, human), drivers and indicators; and
- Block 3: Development of a methodology to baseline the key vulnerabilities and linked indicators.

In 2017 two grant agreements with the Member States were used for the development of methodologies and collaboration tools for emerging risk knowledge exchange (DEMETER) and data collection on ciguatera food poisoning in Europe (EuroCigua). Another project undertaken by EFSA deals with climate changes and food safety (CLEFA).

EFSA is also collaborating with the TNO (an independent Dutch research organization that focuses on applied science) to develop an Emerging Risk Identification Support (ERIS) project that combines the automated selection of relevant information and expert knowledge for the identification of new and emerging issues in the area of food and feed safety. Moreover, a total of 17 potential "emerging issues" were discussed in 2017 and assessed against a set of predefined criteria: (a) new hazard, (b) new or increased exposure, (c) new susceptible groups and (d) new drivers.

11.9 RESEARCH ACTIVITY

EFSA's research coordination objectives include the aims to:

- become knowledge broker for a wider RA community;
- advocate uptake of regulatory research needs;
- build synergies with research projects avoiding duplication;

- early access to deliverables/outcomes from research projects; and
- foster impactful research that feeds regulatory science/policy/ decision-making.

The specific budget under EFSA's science grant and procurement schemes in the period between 2012 and 2016 has been in M€9.43, 11.0, 11.3, 8.83 and 9.77.

The first "Risk Assessment Research Assembly" (RARA) was organized by EFSA in Utrecht on 7 February 2018, with a view of stimulating new partnership in food safety research.

The EFSA Research Platform has been established in response to a call for EFSA becoming a knowledge broker between researchers and policy makers. It offers access to the following tools:

- funding programmes;
- upcoming calls;
- partner search;
- ongoing research; and
- project ideas.

The main research streams include:

- safe food systems:
  - improve food safety while moving towards alternative and sustainable production systems;
- innovation in risk assessment:
  - anticipating impact of innovations and new technologies on integrated risk assessment; and
- holistic risk assessment:
  - understanding the context and delivering and communicating impactful science.

The horizons for food safety research include:

- food 2030;
- strategy papers;
- EU project databases; and
- external projects – EFSA conference 2018 on science, food and society.

The RTD workshop, held on 17 January 2019, discussed the food safety system of the future and the possible needs for action and explored options and opportunities for the possible creation of a European research infrastructure/platform.

A new office in EFSA, science studies and project identification and development (SPIDO), is dealing with science studies and projects. This new function in EFSA aims at supporting, by making use of *ad hoc* financial resources, the development of new methodologies useful for risk assessment. Calls for specific proposals to be supported by EFSA are expected in the next future in the areas of:

- new approaches and methodologies;
- artificial intelligence; and
- multiple chemicals risk assessment.

11.10 RISK COMMUNICATION AND THE *EFSA JOURNAL*

Communication is a fundamental duty of EFSA that is largely delivered through the *EFSA Journal* and a number of other publications and communication tools.

The *EFSA Journal* is an open-access, online scientific journal that publishes the scientific outputs of EFSA. EFSA's various output types are devoted to the field of risk assessment in relation to food and feed and include nutrition, animal health and welfare, plant health and plant protection. The *EFSA Journal* is the single repository of EFSA's scientific outputs that helps to raise visibility and awareness of EFSA's scientific work.

The *EFSA Journal* does not accept contributions from third parties and a classical peer review of EFSA scientific outputs is not carried out because EFSA has very stringent procedures in place to ensure that each output is based on the current status of sound science before final adoption.

EFSA Scientific Outputs and Supporting Publications are as follows:

A. EFSA Scientific Outputs

There are eight types of EFSA's scientific outputs: "Opinions of Scientific Committee/Panel"; and "Other Scientific Outputs". They all are published in the EFSA Journal. The described scientific outputs find their legal basis in Chapter III of the Founding Regulation, specifically in Article 28 to Article 40.

A.1. Scientific Opinions of Scientific Committee/Scientific Panel.

Scientific opinions are prepared and adopted by the Scientific Committee or a Scientific Panel.

A.1.1 Opinion of the Scientific Committee/Scientific Panel

Opinions include for example risk assessments on general scientific issues, evaluations of an application for the authorization of a product, substance or claim or an evaluation of a risk assessment.

A.1.2 Statement of the Scientific Committee /Scientific Panel

A Statement of the Scientific Committee or Scientific Panel is a scientific output in the form of a concise document that does not go into the same level of detail as an opinion.

A.1.3 Guidance of the Scientific Committee/Panel

Guidance of the Scientific Committee or Scientific Panel explains EFSA's procedures and approaches to risk assessments. Guidance documents may also specify the information and data which industry must provide when submitting applications to EFSA for evaluation prior to their authorization by risk managers.

A.2 Other Scientific Outputs of EFSA.

EFSA can issue other scientific outputs prepared by an EFSA working group and/or by EFSA scientific staff. Their content and publication are approved by the executive director of EFSA.

A.2.1 Statement of EFSA

A statement of EFSA is a document addressing an issue of concern and prepared as advice or factual statement prepared normally within a relatively short time frame.

A.2.2 Guidance of EFSA

Guidance of EFSA explains the principles of scientific aspects of the procedural issues or assessment practices and approaches and may also explain scientific guiding principles on the best practices for the monitoring, reporting and analysis of data.

A.2.3 Conclusions (e.g. on Pesticides Peer Review)

An EFSA Conclusion is a comprehensive scientific evaluation of whether the active substance used in a plant protection product is expected to meet the approval criteria, as foreseen in the relevant legislative framework.

A.2.4 Reasoned Opinion

A Reasoned Opinion, as specified in Regulation (EC) 396/2005, describes the comprehensive scientific evaluation and subsequent conclusions on the risk assessment of pesticide residues resulting from the use of pesticides.

A.2.5 Scientific Report of EFSA

A Scientific Report of EFSA is a scientific document whose main purpose is, for example, to describe original

176

research results that pertain for example to a literature review, statistical data analysis or monitoring results.

B. Supporting publications

In addition to the scientific outputs, EFSA can publish supporting publications on its website, which are not published in the EFSA Journal.

B.1 Technical Report

EFSA can issue a Technical Report at the request of the Commission, on its own initiative or, as foreseen in relevant sectors, by legislation. A Technical Report is a document that describes the nature, state of the art, progress or results of a technical process. The content and publication of a Technical Report are approved by the executive director.

B.2 External Scientific Report

An External Scientific Report is a document, not necessarily reflecting EFSA's views, that describes, data collection, literature review or the development of models used in risk assessment. EFSA can request a beneficiary of a grant (in accordance with Article 36 of EFSA's Founding Regulation) or a contractor (in accordance with EU and EFSA public procurement rules) to produce an External Scientific Report. EFSA can also request a joint working group consisting of experts proposed by EFSA's Advisory Forum and Scientific Committee and Panels (so-called ESCO groups) together with staff members to produce an External Scientific Report. The External Scientific Report is accepted by the head of the relevant scientific unit and the relevant director. Publication of the External Report is a decision of the executive director.

B.3 Event Report

EFSA can organize events such as scientific Colloquia, other scientific events or workshops and thereafter issue Event Reports. An Event Report can be prepared by EFSA staff or by a contractor upon request of the executive director. They may contain the presentations given at the event as well as summaries of the events' discussions, outcomes and conclusions. Event Reports can also be available as printed books with an ISBN reference.

11.11 EXAMPLES OF EFSA SCIENTIFIC COMMITTEE ACTIVITIES

Some achievements and challenges of the EFSA Scientific Committee since its inception have been already reviewed by Silano et al. (2012) or addressed in the previous sections of this chapter. The Scientific Committee activities and achievements are many, and in some cases very intensive. The present section only deals with two projects of the Scientific Committee, chosen as examples of its activity.

11.11.1 The Botanical Compendium

The specificities of botanicals include:

- different varieties/cultivar/geographical areas may contain different levels of active substances;
- different plant parts may be characterized by different phyto-chemical compositions;
- phytochemical compounds are characterized by a huge difference in hydrophilic or lipophilic properties and activities;
- many different types of extraction are possible. Moreover, different extraction time even with the same solvent may produce different phytochemical contents;
- very few clinical studies exist and when available, they mostly deal with purified components; and
- botanical substances responsible for adverse or beneficial effects often are not known.

Risk assessment of botanicals and botanical preparations intended for use in food and food supplements has received considerable attention also before 2008 (Silano and Silano, 2006; Rietjens et al., 2008; Spweijers et al., 2009), but it became highly systematic with the activity of EFSA SC on the Compendium of Botanicals, developed since 2009. This is a hazard database containing the following information:

- botanical scientific name and synonyms;
- botanical family;
- plant part(s) containing the substance(s) of concern;
- substance(s) of possible concern on human health, because:
  - belonging to one of the chemical groups considered as "of concern" by default by the *ad hoc* working group; and

- known to the *ad hoc* working group to be of concern for other reasons.

A web-based version of the Compendium of Botanicals has been developed between 2016 and 2019. The current version available online is a work-in-progress version. The next steps are expected in the period 2020–2023. They will include:

- characterization of the toxicity of the large number of substances listed in the EFSA Compendium as of possible concern for human health, partially also covered by the Open Food Tox database.
- systematic review of the scientific literature and consultation of EFSA Partners databases websites (NTP, JECFA, EMA, ECHA etc.), gathering information on toxicity, genotoxicity, mutagenicity and outcome of already performed safety assessments; and
- liaising with data colleagues to ensure that tox data coded for the Compendium are compatible with the Open Food Tox.

11.11.2 EFSA Chemical Hazards Database (Open Food Tox)

For individual substances, a summary of human health and – depending on the relevant legislation and intended uses – animal health and ecological hazard assessments have been collected and structured into EFSA's chemical hazards database, denominated Open Food Tox. This database provides open-source data for the substance characterization, the links to EFSA's related output, European background legislation and a summary of the critical toxicological endpoints and reference values for a large number of food ingredients, pesticides, contaminants, food contact materials and feeds.

The adoption of the open-data approach by EFSA implies:

- the improvement of interoperability to facilitate a data exchange; and
- the migration towards structured scientific data.

The purposes of the Open Food Tox include:

- inventory of EFSA's chemical risk assessment since its establishment in 2002;
- easy reference and crisis:
  - crisis: free, quick, easy access to EFSA's Chemical Hazards Data; and
  - tool for stakeholders.

- international harmonization and data sharing:
  - OECD harmonized templates; and
  - data sharing.

The Open Food Tox & EFSA content includes:

- chemical information: substance identity;
  - CAS, IUPAC, SMILES, etc.;
  - single substances (e.g. flavourings); and
  - group of substances (e.g. mixture or formulation).
- EFSA outputs:
  - title, publication date, link, etc.;
  - opinions;
  - conclusions on pesticides; and
  - statements.
- toxicological information:
  - genotoxicity;
  - reference points;
  - human and animal health;
  - ecological RA;
  - reference values;
  - regulated products, e.g. ADI for pesticides;
  - nutrients: e.g. DRV for vitamins and minerals; and
  - contaminants: TDI for acrylamide

The future of the Open Food Tox includes (between 2018–2022 a very large increase of the number of substances addressed.

11.12 THE EFSA PERFORMANCE REPORT AND THE CONSOLIDATED ANNUAL ACTIVITY REPORT

The performance report contains data and analyses that make possible to evaluate the outcomes related to the performance of EFSA towards the results expected for the strategic objectives. They are measured through the intermediate impact indicators, outcome indicators and output indicators that were included in the results-based approach model implemented in EFSA in 2017. The EFSA performance at a glance report for each year is available in the "Consolidated Annual Activity Report", thus making possible a good understanding of the quality and extent of EFSA's activities and results.

EFSA produces three performance reports per year:

- the Performance Report (P1), analyzing data as of the end of April;
- the Performance Report (P2): analyzing data as of the end of August; and
- the Performance Report (P3), analyzing data as of the end of October.

The first Performance Report of EFSA is designed to include information in all aspects of the EFSA performance measured during the respective period of reporting against the multiannual work programme as presented in the Programming Document 2019–2021 adopted in December 2018.

EFSA's Performance Indicators set in the Programming Document 2017–2019 are structured in three levels:

- intermediate impact (result) indicators measure EFSA's performance to achieve its strategic objectives in the long-term from all related activities in a strategic area. Because of their nature, most of these indicators are measured on annual or multiannual basis;
- outcome (result) indicators measure the result of several activities towards achieving partially a strategic objective in the midterm. These indicators are measured less frequently, and information on these are included in the report only if the frequency of measurement is relevant to the quarter; and
- output indicators are the annual work program indicators directly linked to the outputs of specific activities in a strategic area (processes or process improvement initiatives or projects) during the year.

In 2018, in terms of compliance with the scientific production deadlines, EFSA registered good results in all the strategic objectives, with the one exception of timeliness in the area of regulated products (83.6% of the outputs were closed on time against a target of 90%, but this delay was limited to few areas where a significantly high workload was experienced). The human resources and the budget employed in the production of scientific opinions were in line with the plans for 2018, and also with the introduction of a flexible resource management approach to tackle the work peaks, leading to a shift of resources mid-year to the area of novel foods. Finally, it is worth highlighting that EFSA fully respected its commitment to scientific independence, as all DOIs of EFSA's scientific experts were submitted in compliance with the current rules.

The most recent Performance Report summarizes the progress achieved in the first reporting period (P1), measured through the performance indicators (cumulative numbers for the period January–April 2019) and compared to the annual targets set in the EFSA Programming Document 2019–2021 (split per reporting period).

The overall performance for the reporting period 2019–until the end of April is satisfactory, with the majority of the indicators measured in the period reaching their targets. The results are consistent across all the 5 EFSA's Strategic Objectives, previously described in this chapter. On average, the ratio of the indicators meeting or exceeding the target stands at 86%, a result higher than the once achieved in P1 2018 (when it was 80%). Out of a total of 72 indicators that are considered relevant for this reporting period, only three registered a "relevant deviation" from their quarterly targets and are closely monitored. The remaining 60 indicators were not measured in this reporting period, either because their first measurement is still pending (12 indicators) or because the measurement does not occur on a quarterly basis (48 indicators). It should also be mentioned that, every six years, an external contractor is tasked to assess EFSA activities in its scientific, communication and engagement work and that the report of the third independent evaluation of EFSA recognized its many achievements. This report, made available in its final format on 18 June 2018, shows that EFSA has made significant progress in addressing weaknesses previously identified. Between 2011 and 2016, EFSA's mechanisms for cooperation and engagement with partners and stakeholders at national, EU and international level were strengthened, contributing to enhanced risk assessment capacity at the EU level. In response to demands for greater transparency and a need to maintain trust, EFSA has committed to reinforcing and refocusing efforts on transparency and independence. Crucially, EFSA has now strengthened its independence policy and rules and set out a plan to move towards an "Open Science Organization", through its "Transparency and Engagement in Risk Assessment" project. Linked to this, EFSA improved mechanisms for engagement with stakeholders and cross-cutting communication activities have contributed to improved clarity, accessibility and professionalism of its materials.

The EFSA Consolidated Annual Activity Report is a very useful document for all those interested to know in details the work carried out by EFSA in a specific year. For instance, the report for the year 2018 provides, in addition to the performance at a glance, detailed information on the achievements of the work program to:

- prioritize public and stakeholder engagement in the process of scientific assessment;
- widen EFSA's evidence base and optimize access to its data;
- build the EU's scientific assessment capacity and knowledge community;
- prepare for future risk assessment challenges; and
- create an environment and culture that reflects EFSA's values.

Moreover, this report covered the management of resources and the assurance, dealing with:

- budget and finance management;
- human resources management;
- full-time equivalents (FTEs) and budget indicators;
- assessment of audit results during the reporting year;
- management assurance-10 assurance pillars;
- declaration of assurance;
- statement of the head of business services; and
- management board assessment.

Finally, seven annexes cover the following subjects:

- Annex I – resource allocations for per strategic objective;
- Annex II – financial resources;
- Annex III – the status of projects;
- Annex IV – questions closed 2018;
- Annex V – human resources;
- Annex VI – negotiated procedures and time to grant; and
- Annex VII – annual report on the implementation of EFSA's policy on independence – reporting period: 1 January–31 December 2018.

12

The New Regulation (EU) 2019/1381 on the Transparency and Sustainability of the EU Risk Assessment in the Food Chain

12.1 INTRODUCTION

In September 2019, Regulation (EU) 2019/1381 was adopted by the European Parliament and the Council on the transparency and sustainability of the EU risk assessment in the food chain. This Regulation that amends Regulations (EC) 178/2002, (EC) 1829/2003, (EC) 1831/2003, (EC) 2065/2003, (EC) 1935/2004, (EC) 1331/2008, (EC) 1107/2009, (EU) 2015/2283 and Directive 2001/18/EC enters into force on 27 March 2021, whereas specific sections (i.e. Articles 1(2) and 1(3)) will apply later.

Regulation (EU) 2019/1381 aims at:

- improving and clarifying the rules on transparency, especially with regard to the scientific studies supporting the risk assessment, also through the establishment of the Union Register of Studies;

- increasing the reliability, objectivity and independence of studies used by EFSA in its risk assessment, in particular in the framework of authorization dossiers, also through the adoption of verification studies;
- improving the governance and strengthening the scientific cooperation and involvement of the Member States in EFSA work;
- addressing the limitations affecting the long-term scientific capacity of EFSA and its ability to maintain a high level of scientific expertise across the different areas of the agri-food sector, also taking into account the related financial and budgetary aspects; and
- developing a more effective and transparent risk communication with the public in collaboration with the Member States, addressing the objectives, the general principles and the general plan.

Several consultations were organized in 2018 to discuss the preparation of this new Regulation:

- Open Public Consultation of the Commission's Roadmap;
- Open Public Consultation via a Questionnaire;
- Advisory Group – on the Food Chain, Animal and Plant Health – All the Main Stakeholders;
- EFSA's Advisory Forum – National Food Safety Bodies;
- EFSA's Scientific Committee – Scientists; and
- Expert Group on General Food Law – Member States.

Moreover, the Commission's proposal on the transparency and sustainability of the EU risk assessment in the food chain, proposing targeted changes to the EU General Food Law and eight related sectoral legislative acts, was presented to the Member States in the AGRIFISH Council on 16 April 2018 and to the ENVI Committee of the European Parliament on 26 April 2018.

Regulation (EU) 2019/1381 on the transparency and sustainability of EU risk assessment in the food chain has been adopted by the European Parliament and the Council and has been published in *Official Journal of the European Union* (OJEU) on 6 September 2019. The intent of the new Regulation is to increase the transparency of risk assessments in the food chain in the EU and to strengthen the reliability, objectivity and independence of the studies used by EFSA.

The main elements of this new Regulation aim at the following objectives:

- **Ensuring more transparency**: citizens will have automatic access to all studies and information submitted by industry in the risk assessment process. Stakeholders and the general public will also be consulted on submitted studies. At the same time, the

186

Regulation will guarantee confidentiality, in duly justified circumstances, by setting out the type of information that may be considered significantly harmful for commercial interests and therefore cannot be disclosed.

- **Increasing the independence of studies**: EFSA will be notified of all commissioned studies to guarantee that companies applying for authorizations submit all relevant information and do not hold back unfavourable studies. The authority will also provide general advice to applicants, prior to the submission of the dossier.
- **Strengthening the governance and the scientific cooperation**: The Member States, civil society and European Parliament will be involved in the governance of EFSA by being duly represented in its management board. The Member States will foster the EFSA's scientific capacity and engage the best independent experts in its work.
- **Developing comprehensive risk communication**: a general plan for risk communication will be adopted and will ensure a coherent risk communication strategy throughout the risk analysis process, combined with open dialogue amongst all interested parties.

Under the new rules, studies and information supporting a request for a scientific output by EFSA are to be made public automatically when an application by an FBO is validated or found admissible. Confidential information will be protected in duly justified circumstances, and confidentiality requests will be assessed by EFSA.

Other measures introduced by the revamped General Food Law include:

- the possibility for the European Commission to ask EFSA to commission additional studies for verification purposes of the evidence used in its risk assessment and to organize fact-finding missions to verify the compliance of laboratories/studies with relevant standards;
- a new database of studies commissioned by FBOs; and
- a more active role for the Member States in helping EFSA attracting more and the best scientists to participate in scientific panels.

As reported by the European Commission – DG SANTE Units D1 and E1 (2018) – there are four main pillars in this Regulation:

1. transparency of risk assessment;
   - studies/data-supporting applications for authorization to be made public proactively and early on in the risk assessment process, except for confidential info;

- lists of the info that may be considered confidential; and
- no prejudice to intellectual property rights and data exclusivity.
2. quality and reliability of studies
   - EU register of commissioned studies;
   - pre-submission procedure:
     - consultation on planned studies (renewals only) and submitted studies (renewals and new requests);
     - commissioning of verification studies in exceptional cases; and
     - control and audit of laboratories carrying out studies.

3. sustainability and governance of the risk assessment system
   - alignment with the common approach on decentralized agencies (MS representatives in the management board); and
   - MS involved in the Panel Members' selection process: the pool of experts (criteria of excellence and independence to be respected).
4. improved risk communication
   - definition of precise objectives;
   - general principles; and
   - general communication plan.

If requested, EFSA will provide pre-submission advice on the relevant provisions and required content of a petition or notification. However, the EFSA staff providing pre-submission advice will not be involved in any preparatory scientific or technical work directly or indirectly relevant to the application or notification. A summary of pre-submission advice will be made public once the application is considered valid or admissible.

12.2 THE MAIN PROCEDURAL INNOVATIONS

12.2.1 Risk Communication

Article 1 of Regulation (EU) 2019/1381 amends Regulation (EU) 178/2002 by inserting in Chapter II, Section 1a, dealing with risk communication, consisting of two articles: Article 8a on the objectives of risk communication and Article 8b on general principles of risk communication. Moreover, a general plan for risk communication must be adopted and systematically updated by the European Commission. Risk communication shall pursue the following objectives:

(a) raise awareness and understanding of the specific issues;
(b) ensure consistency, transparency and clarity in formulating risk management and provide a sound basis for understanding risk management decisions;
(c) improve the overall effectiveness and efficiency of the risk analysis and foster public understanding of the risk analysis;
(d) ensure appropriate involvement of consumers, feed and food businesses, the academic community and all other interested parties; and
(e) contribute to the fight against the dissemination of false information and the sources thereof and to ensure the provision of information to consumers about risk prevention strategies.

The general principles of risk communication should:

(a) ensure that all appropriate information is exchanged in an interactive, accurate and timely manner with all interested parties, based on the principles of transparency, openness, and responsiveness;
(b) provide transparent information at each stage of the risk analysis process by taking into account risk perceptions of all interested parties and facilitating understanding and dialogue amongst all interested parties; and
(c) be clear and accessible, including to those not directly involved in the process or not having a scientific background, while duly respecting the applicable legal provisions on confidentiality and protection of personal data.

Moreover, according to Article 8c, the European Commission has been charged with the adoption, by means of implementing acts, of a general plan for risk communication in order to achieve the objectives set out in Article 8a, in accordance with the general principles set out in Article 8b. The Commission shall adopt the general plan for risk communication within two years from the application of this Regulation and shall keep it updated, taking into account technical and scientific progresses and experience gained.

12.2.2 Membership of the EFSA Management Board

Another very important amendment concerns Article 25 of Regulation (EC) 178/2002 whose paragraph 1 has been modified with the objective of ensuring the participation in the EFSA management board of representatives of the Member States and the European Institutions as well as from civil society and food chain interest.

To this end, each Member State shall nominate a member and an alternate member as its representatives to the EFSA Management Board. The members and alternate members thus nominated shall be appointed by the Council and have the right to vote.

> In addition to members and alternate members referred to in the above the EFSA Management Board shall include:
>
> (a) two members and two alternate members appointed by the Commission as its representatives, with the right to vote;
> (b) two members appointed by the European Parliament, with the right to vote;
> (c) four members and four alternate members with the right to vote as representatives of civil society and food chain interests, namely one member and one alternate member from consumer organizations, one member and one alternate member from environmental non governmental organizations, one member and one alternate member from farmers organizations, and one member and one alternate member from industry organizations.

The members of the Management Board shall be appointed for a period of four years, taking into account high competence in the area of food safety risk assessment.

12.2.3 The Pre-Submission Advice Procedure

The new "pre-submission advice procedure" has been introduced by Article 32a of Regulation (EU) 2019/1381, to make possible such a procedure avoiding any possible overlap with the submission procedure. This article has stated that:

> Where Union law contains provisions for EFSA to provide a scientific output, including a scientific opinion, the staff of EFSA shall, at the request of a potential applicant or notifier, provide advice on the rules applicable to, and the content required for, the application or notification, prior to its submission. Such advice provided by the staff of EFSA shall be without prejudice and non-committal as to any subsequent assessment of applications or notifications by the Scientific Panels. The EFSA staff providing the advice shall not be involved in any preparatory scientific or technical work that is directly or indirectly relevant to the application or notification that is the subject of the advice.

The Authority shall publish general guidance on its website regarding the rules applicable to, and the content required for, applications and notifications, including, where appropriate, general guidance on the design of required studies.

12.2.4 The Union Register of Studies

The procedure for the notification of studies has been addressed by the new Regulation in Article 32b. The text of the article 32b of Regulation (EU) 2019/1381 is as follow:

EFSA shall establish and manage a database of studies commissioned or carried out by business operators to support an application or notification in relation to which Union law contains provisions for EFSA to provide a scientific output, including a scientific opinion.

For the purposes of paragraph 1, business operators shall, without delay, notify EFSA of the title and the scope of any study commissioned or carried out by them to support an application or a notification, as well as the laboratory or testing facility carrying out that study, and its starting and planned completion dates. For the purposes of paragraph 1, laboratories and other testing facilities located in the Union shall also, without delay, notify EFSA of the title and the scope of any study commissioned by business operators and carried out by such laboratories or other testing facilities to support an application or a notification, its starting and planned completion dates, as well as the name of the business operator who commissioned such a study. This paragraph shall also apply, mutatis mutandis, to laboratories and other testing facilities located in third countries insofar as set out in relevant agreements and arrangements with those third countries, including as referred to in Article 49.

An application or notification shall not be considered valid or admissible where it is supported by studies that have not been previously notified, unless the applicant or notifier provides a valid justification for the non-notification of such studies.

Where studies have not been previously notified in accordance with paragraph 2 or 3, and where a valid justification has not been provided, an application or notification may be re-submitted, provided that the applicant or notifier notifies to EFSA those studies, in particular their title and their scope, the laboratory or testing facility carrying them out as well as their starting and planned completion dates.

The assessment of the validity or the admissibility of such re-submitted application or notification shall commence six months after the notification of the studies pursuant to the second subparagraph.

An application or notification shall not be considered valid or admissible, where studies that have previously been notified in accordance with paragraph 2 or 3 are not included in the application or notification, unless the applicant or notifier provides a valid justification for the non-inclusion of such studies.

Where the studies which have previously been notified in accordance with paragraph 2 or 3 were not included in the application or notification, and where a valid justification has not been provided, an application or notification may be resubmitted, provided that the applicant or notifier submits all the studies that were notified in accordance with paragraph 2 or 3.

The assessment of the validity or admissibility of such re-submitted application or notification shall commence six months after the submission of the studies pursuant to the second subparagraph.

Where EFSA detects, during its risk assessment, that studies notified in accordance with paragraphs 2 or 3 are not included in the corresponding application or notification in full, and in the absence of a valid justification of the applicant or notifier to that effect, the applicable time limits within which EFSA is required to deliver its scientific output shall be suspended. That suspension shall end six months after the submission of all data of those studies.

EFSA shall make public the notified information only in cases where it received a corresponding application or notification and after EFSA has decided on the disclosure of the accompanying studies in accordance with Articles 38 to 39e.

EFSA shall lay down the practical arrangements for implementing the provisions of this Article, including arrangements for requesting and making public the valid justifications in the cases referred to in paragraphs 4, 5 and 6. Those arrangements shall be in accordance with this Regulation and other relevant Union Law.

12.2.5 Consultation of Third Parties

Consultation of third parties will take place according to Article 32c of Regulation (EU) 2019/1381 stating that:

Where the relevant Union law provides that an approval or an authorization, including by means of a notification, may be renewed, the potential applicant or notifier for the renewal shall notify EFSA of the studies it intends to perform for that purpose, including information on how the various studies are to be carried out to ensure compliance with regulatory requirements. Following such notification of studies, EFSA shall launch a consultation of stakeholders and the public on the intended studies for renewal, including on the proposed design of studies. Taking into account the received comments from the stakeholders and the public which are relevant for the risk assessment of the intended renewal, EFSA shall provide advice on the content of the intended renewal application or notification, as well as on the design of the studies. The advice provided by EFSA shall be without prejudice and non-committal as to the subsequent assessment of the applications or notifications for renewal by the Scientific Panels.

EFSA shall consult stakeholders and the public on the basis of the non confidential version of the application or notification made public by EFSA in accordance with Articles 38 to 39e, and immediately after such disclosure to the public, in order to identify whether other relevant scientific data or studies are available on the subject matter concerned by the application or notification. In duly justified cases, where there is a risk that the results of the public consultation performed in accordance with this paragraph cannot be given proper consideration because of the applicable time limits within which EFSA is required to deliver its scientific output, those time limits may be extended for a maximum period of seven weeks. This paragraph is without prejudice to the EFSA's obligations under Article 33 and does not apply to the submission of any supplementary information by the applicants or notifiers during the risk assessment process.

EFSA shall lay down the practical arrangements for implementing the procedures referred to in this Article and Article 32a.

12.2.6 Verification Studies

According to Article 32d of Regulation (EU) 2019/1381, without prejudice to the obligation on applicants to demonstrate the safety of a subject matter submitted to a system of authorization, the European Commission, in exceptional circumstances of serious controversies or conflicting results, may request EFSA to commission scientific studies with the objective of verifying evidence used in its risk assessment process (verification studies). The studies commissioned may have a wider scope than the evidence subject to verification.

12.2.7 Confidentiality

The rules concerning confidentiality are addressed in Article 39 of Regulation (EU) 2019/1381 that states:

> By way of derogation from Article 38, EFSA shall not make public any information for which confidential treatment has been requested under the conditions laid down in this Article.

Upon the request of an applicant, EFSA may grant confidential treatment only with respect to the following items of information where the disclosure of such information is demonstrated by the applicant to potentially harm its interests to a significant degree:

(a) the manufacturing or production process, including the method and innovative aspects thereof, as well as other technical and industrial specifications inherent to that process or method, except for information which is relevant to the assessment of safety;

(b) commercial links between a producer or importer and the applicant or the authorization holder, where applicable;

(c) commercial information revealing sourcing, market shares or business strategy of the applicant; and

(d) quantitative composition of the subject matter of the request, except for information which is relevant to the assessment of safety.

The list of information referred to in paragraph 2 shall be without prejudice to any sectoral Union law.

Notwithstanding paragraphs 2 and 3:

(a) where urgent action is essential to protect human health, animal health or the environment, such as in emergency situations, EFSA may disclose the information referred to in paragraphs 2 and 3;

(b) information which forms part of conclusions of scientific outputs, including scientific opinions, delivered by EFSA and which relate to foreseeable effects on human health, animal health or the environment, shall nevertheless be made public.

The specific rules and procedures relevant in the sector of confidentiality have been addressed in Article 39a (Request for confidentiality); Article 39b (Decision for confidentiality); Article 39c (Review of confidentiality); and Article 39d (Obligations with regard to confidentiality).

12.2.8 Protection of Personal Data

Modalities for the protection of personal data are considered in Article 39e of Regulation (EU) 2019/1381 that states:

With respect to requests for scientific outputs, including scientific opinions under Union law, the Authority shall always make public:

(a) the name and address of the applicant;

(b) the names of authors of published or publicly available studies supporting such requests; and

(c) the names of all participants and observers in meetings of the Scientific Committee and the Scientific Panels, their working groups and any other ad hoc group meeting on the subject matter. Notwithstanding paragraph 1, disclosure of names and addresses of natural persons involved in testing on vertebrate animals or in obtaining toxicological information shall be deemed to significantly harm the privacy and the integrity of those natural persons and shall not be made publicly available unless otherwise specified in Regulations (EU) 2016/679 and (EU) 2018/1725 of the European Parliament and of the Council.

Regulations (EU) 2016/679 and (EU) 2018/1725 shall apply to the processing of personal data carried out pursuant to this Regulation. Any personal data made public pursuant to Article 38 of this Regulation and this Article shall only be used to ensure the transparency of the risk assessment under this Regulation and shall not be further processed in a manner that is incompatible with these purposes, in accordance with point (b) of Article 5(1) of Regulation (EU) 2016/679 and point (b) of Article 4(1) of Regulation (EU) 2018/1725, as the case may be.

12.2.9 Standard Data Formats

According to Article 39f, in order to ensure the efficient processing of requests to EFSA for a scientific output, standard data formats shall be adopted in accordance with paragraph 2 of this Article to allow documents to be submitted, searched, copied and printed, while ensuring compliance with regulatory requirements set out in Union law.

Those standard data formats shall:
(a) not be based on proprietary standards;
(b) ensure interoperability with existing data submission approaches to the extent possible;
(c) be user-friendly and adapted for the use by small and medium-sized enterprises.

For the adoption of standard data formats referred to in paragraph 1, the following procedure shall be followed:
(a) EFSA shall draw up draft standard data formats for the purposes of the different authorization procedures and relevant requests for a scientific output by the European Parliament, by the Commission and by the Member States;
(b) the Commission shall, taking into account the applicable requirements in the different authorization procedures and other legal frameworks and following any necessary adaptations, adopt,

195

by means of implementing acts, standard data formats. Those implementing acts shall be adopted in accordance with the procedure referred to in Article 58(2);

(c) EFSA shall make the standard data formats, as adopted, available on its website;

(d) where standard data formats have been adopted pursuant to this Article, applications as well as requests for a scientific output, including a scientific opinion by the European Parliament, by the Commission and by the Member States, shall only be submitted in accordance with those standard data formats.

12.2.10 Information Systems

According to Article 39g, the information systems operated by EFSA to store its data, including confidential and personal data, shall be designed in a way that guarantees that any access to it is fully auditable and that the highest standards of security appropriate to the security risks at stake are attained, taking into account Articles 39a to 39f of Regulation (EU) 2019/1381.

12.2.11 Exercise of the Delegation

Before adopting a delegated act, the Commission has to consult experts designated by each Member State in accordance with the principles laid down in the Inter-Institutional Agreement on Better Law-Making of 13 April 2016. As soon as it adopts a delegated act, the Commission shall notify it simultaneously to the European Parliament and to the Council. The delegated act shall enter into force only if both the two notified European Institutions will let known within a period of two months from the notification that they do not object. This two-month period may be extended to four months by the initiative of the Parliament or Council.

12.2.12 Review Clause

The European Commission shall ensure the regular review of the application of this Regulation. Every five years, the Commission shall assess EFSA's performance in relation to its objectives, mandate, task procedures and location, in accordance with Commission guidelines. The evaluation shall address the possible need to modify the EFSA's mandate and the financial implications of any such modifications. Many other significant improvements have been adopted by Regulation (EU) 2019/1381. This is the case, for instance, of the duration of the mandate of the members of scientific panels and scientific committee or the procedure for their designations by the Member States.

12.3 FUTURE CHALLENGES FOR EFSA AND OTHER EUROPEAN INSTITUTIONS

The adoption of the new EU Regulation on the transparency and sustainability of the EU risk assessment in the food chain demands considerable efforts from EFSA and other relevant institutions and, in particular:

- readiness;
  - implementation of the new requirements from the European Commission;
  - closure of critical gaps in EFSA Strategy 2020 and completion the updating of this strategy for the period 2021–2027;
- efficiency and efficacy
  - enabling new processes; and
  - keeping a high quality of science;
- digitalization
  - lifting risk assessment methods to adapt to an external environment and science that is increasingly digital and social.

The priorities determined by the new conditions applicable to EFSA include:

- **transparency and confidentiality**
  - increased transparency through the establishment of a register of studies and the proactive publication studies results;
  - centralized confidentiality checks; and
  - clarification of interplay between public access to documents and Aarhus Regulations.
- **engagement and risk communication**
  - inclusive risk communication (EC, MSs); and
  - improving coordination between risk assessors and risk managers to ensure better communication to stakeholders and the general public.
- **scientific value**
  - more active role of the Member States in building EFSA's pool of experts;
  - stronger emphasis on experts' involvement in active research;

- – enhanced flexibility for EFSA scientific production, including its staff;
- – strengthened scientific support to applicants and small/medium enterprises (pre-submission advice);
- – improved design and coherence of studies underlying EFSA scientific outputs; and
- – enlarged evidence base of EFSA scientific assessment.
- **governance**
  - – Member States, European Parliament, and stakeholders' representatives in the management board;
  - – broader expertise of management board members; and
  - – more prominent role for interagency cooperation.

13

Better Training for Safer Food

13.1 LEGAL BASIS

Article 51 of Regulation (EC) 882/2004 on official controls on food and feed, animal health and animal welfare rules and Article 2 of Council Directive (EC) 2000/29 on protective measures on plants or plant products provide the legal basis for this initiative. They empower the Commission to develop training for competent authority staff in EU countries and non-EU countries responsible for controls in the covered areas.

13.2 MAIN OBJECTIVES OF TRAINING

The main objectives of the initiative Better Training for Safer Food are the organization and development of an EU training strategy with a view to:

- ensuring and maintaining a high level of consumer protection and animal health, animal welfare and plant health;
- improving and harmonizing official controls in EU countries and creating the conditions for a level-playing field for food businesses contributing to EU priority on jobs and growth;
- ensuring the safety of food imports from non-EU countries on the EU market and ultimately to reducing risks for EU consumers and providing EU businesses with easier access to safe goods from non-EU countries;

- ensuring harmonization of control procedures between EU and non-EU partners in order to guarantee a parallel competitive position of EU businesses with their non-EU counterparts;
- building confidence in the EU regulatory model with competent authorities of other international trade partners and pave the way for new food market opportunities and increased competitiveness for EU operators; and
- ensuring fair trade with non-EU countries and in particular developing countries.

13.3 DEVELOPMENT OF TRAINING

Training activities have been launched since 2006. The courses are delivered in EU and non-EU countries, targeting the staff of competent authorities dealing with food and feed safety from the EU Member States and selected non-EU countries.

The training is provided mainly through workshops and/or secondment of experts to carry out targeted tailor-made training under various formats. In addition, to reach the maximum of people involved in controls, e-learning modules have been developed since 2010.

The initiative follows the "train the trainers" principle, and participants should disseminate the knowledge acquired from the training among their colleagues in their home countries.

These activities are implemented by external contractors designated through public procurement procedures. They are responsible for designing the training programmes, selecting the best tutors and establishing the training calendar for each activity.

Participation in BTSF activities is channelled through designated competent authorities in each country. BTSF contact points have been appointed in the EU Member States and certain other non-EU countries to coordinate participant selection.

The Commission sets policy and general strategy for Better Training for Safer Food, and the executive agency CHAFEA manages all phases of projects.

For the better training for safer food lecture courses planned to take place in the period 2019-2021 see Table 13.1.

Table 13.1 Lecture Courses for the Better Training for Safer Food Planned to Take Place in the Years 2019, 2020 and 2021

Courses in lecture-halls:
1. Animal nutrition
2. Animal Welfare
3. Animal disease preparedness, including early warning, contingency planning and animal disease control
4. Antimicorobial resistance - phase 2
5. Audit Systems and internal auditing
6. Controls on contaminants in food and feed – Phase 2
7. Evaluation and authorisation procedures for Plant Protection Products. 2019–2020
8. Food additives
9. Food Contact Materials
10. Food labelling and claims, food supplements, foods with added vitamins and minerals and foods for specific groups – phase 2 – 2020/21
11. Foodborne outbreak management – phase 2 – 2019/21
12. Food Hygiene and Flexibility
13. Food hygiene at primary production
14. Inspection and calibration of plant protection product application equipment in compliance with the provisions of Directive 2019/128/EC
15. Integrated Pest Management
16. Microbiological criteria – course zoonoses
17. Movements of dogs and cats
18. New Investigation Techniques of food fraud
19. Organic farming
20. Prevention, Control and eradication of Transmissible Spongiform Encephalopathies (TSE) and By-Products of food origin
21. Prevention and Control of Antimicrobial Resistance – 2nd Phase 2019-2020
22. Protected designations schemes (PDS) – Geographical Indications
23. Plant health surveys – 2nd Phase
24. Plant Health Control 2020/21
25. Risk Assessment in Microbiology
26. Strengthening of official controls
27. Use of the EU trade control and expert system (Traces)
28. Training on auditing general hygiene requirements and control procedures based on the HACCP principles developed by food business operators
29. The use of the EU trade control and expert system – Phase 2 (Traces) 2019–2021
30. Veterinary and food safety import controls in border inspection post
31. Workshop Biocides
32. Workshop on African Swine Fever surveillance and wildlife management e-learning:

(Continued)

Table 13.1 (Continued)   Lecture Courses for the Better Training for Safer Food Planned to Take Place in the Years 2019, 2020 and 2021

e-learing courses:
1. Animal nutrition
2. Animal Welfare at slaughter and killing for disease control
3. Animal Welfare at slaughter and killing for disease control for poultry
4. Animal Health prevention and controls for aquaculture animals
5. Food Contact Materials rules
6. Food hygiene and control on fishery products and live bivalve molluscs
7. EU plant quarantine regime for imports
8. Principle of Hazard Analysis and Critical Control Point audits (HACCP)
9. Prevention, control and eradication of Transmissible Spongiform Encephalopathies (TSE)
10. Rapid Alert System for Food and Feed (RASFF)

Source: Website Italian Ministry of Health

14

Overall Conclusions

When evaluating the impact of new legislations relevant for food safety, it is important to consider that a constant character of these legislations is that, although they enter into force within a few weeks from their publication in *Official Journal*, normally they become applicable only after a certain period (e.g. a couple of years) from their entry into force to provide the time required for the market and institutional transformations imposed by the new Regulations.

Since more than 20 years the European Institutions (i.e. Commission, Council and Parliament) have undertaken an intensive collaboration to improve the standards of food safety in the European Union. Major initiatives and tools of this complex and fundamental programme, announced in the year 2000 with the "White Paper on Food Safety" and analyzed in their levels of implementation in this book, include:

- the establishment of the European Food Safety Authority based on the principles of independence, scientific excellence and transparency of its operations;
- the implementation of an integrated legislative reform to improve and bring coherence in the Community's legislation covering all aspects of food products from "farm to table" with a large number of separate actions addressing, among others, (i) responsibilities of feed manufacturers, farmers and food operators; (ii) traceability of feed, food and its ingredients; (iii) proper risk analysis through (a) risk assessment (scientific advice and information analysis), (b) risk management (regulation and control) and (c) risk communication; and (iv) the application of the precautionary principle as appropriate;

- the Control, as a joint task for the Commission and the Member States, on the implementation of legislation through three core elements: (i) operational criteria set up at the Community level, (ii) Community control guidelines, and (iii) enhanced administrative cooperation in the development and operation of control; and
- the mandatory and voluntary consumer's information that has been considered fundamental to encourage their deeper involvement in food safety practices to better control food-related health risks and emerging food safety concerns.

In 2019, at about 20 years of distance from the start of this successful process, the European Institutions confirmed their interest for achieving and maintaining the highest standards of food safety in the EU by adopting a new Regulation to further reinforce and potentiate EFSA risk assessment in spite of the excellent results already achieved so far as shown by EFSA performance at a glance. On the other hand, it should be clear that the debate on the scientific advice to the European policy is very active, as is shown by the opinion on "Scientific Advice to European Policy in a Complex World" that was delivered to the European Commission by the group of chief scientific advisors in September 2019. The scientific advisors were supported by the SAPEA consortium, which developed a review report supporting the opinion (Making Sense of Science for Policy under Conditions of Complexity and Uncertainty). The SAPEA (Science Advice for Policy by European Academies) consortium brings together knowledge and expertise from more than 100 academies and learning societies in over 40 countries across Europe. Funded through the Horizon 2020 programme of the EU, the SAPEA consortium comprises Academia Europaea; All European Academies; the European Academies Science Advisory Council; the European Council of Academies of Applied Sciences, Technologies and Engineering; and the Federation of European Academies of Medicine.

It is interesting to consider the recommendations provided to the European Commission in this scientific opinion. They were as follows:

Recommendation 1: engage early and regularly.
- clarify boundaries between science, scientific advice and politics; and
- define together the questions for scientific advice.

Recommendation 2: ensure the quality of scientific evidence.
- use the full scope of good science;
- ensure the rigorous synthesis of scientific evidence;

- refine the approach to conflicts of interest; and
- codify good scientific advice and consider oversight of its implementation.

Recommendation 3: analyze, assess and communicate uncertainties.

- use the most suitable uncertainty analysis approaches;
- communicate uncertainties and diverging scientific opinions; and
- explain the path from evidence to advice.

APPENDIX I – RESOLUTION ADOPTED BY THE EUROPEAN PARLIAMENT IN MAY 2016 ON MANDATORY INDICATIONS OF THE COUNTRY OF ORIGIN OR PLACE OF PROVENANCE FOR CERTAIN FOODS

The European Parliament,

--------------------------omissis----------------------------

Drinking milk and milk used as an ingredient in dairy products

1. Points out that Recital 32 of the Food Information to Consumers Regulation states that milk is one of the products for which an indication of origin is considered of particular interest;

2. Emphasises that, according to the Eurobarometer survey 2013, 84% of EU citizens consider it necessary to indicate the origin of milk, whether sold as such or used as an ingredient in dairy products; notes that this is one of several factors that may influence consumer behaviour;

3. Points out that the mandatory indication of the origin of milk, sold as such or used as an ingredient in dairy products, is a useful measure to protect the quality of dairy products and protect employment in a sector which is going through a severe crisis;

4. Notes that, according to the survey accompanying the Commission's report on milk and other meat, the costs of mandatory origin labelling for milk and milk used as an ingredient increase as the complexity of the production process grows; notes that the same survey suggests that businesses in certain Member States had overstated the impact of mandatory origin labelling on their competitive position, as the survey could find no clear explanation for the high cost estimates given by such businesses, but stated that it may be a signal of strong opposition per se to origin labelling;

5. Calls for the establishment of a Commission Working Group to further evaluate the Commission's report, published on 20 May 2015, in order to determine which costs can be reduced to an acceptable level if further mandatory country of origin labelling proposals are limited to dairy and lightly processed dairy products;

6. Appreciates the survey's analysis of the costs and benefits of the introduction of mandatory origin labelling for milk and milk used as an ingredient, but considers that the Commission in its conclusions does not sufficiently take into account the positive aspects of country of origin labelling for such products, such as greater consumer information; notes that consumers can feel misled when information on mandatory origin labelling is not available and other food labels, such as national flags, are used;

7. Stresses the importance of small and medium-sized enterprises in the processing chain;

8. Takes the view that the Commission should take into account and analyze the economic impact of compulsory origin labelling on SMEs in the agricultural and food sectors concerned;

9. Considers that the Commission's conclusion in relation to milk and milk used as an ingredient possibly overstates the costs of country of origin labelling to business as all dairy products are considered together;

10. Notes that the Commission concludes that the costs of country of origin labelling for milk would be modest;

Other types of meat

11. Stresses that, according to the Eurobarometer survey 2013, 88% of EU citizens consider it necessary to indicate the origin of meat other than beef, swine, sheep, goat and poultry meat;
12. Notes that the horsemeat scandal showed the need for greater transparency in the horsemeat supply chain;
13. Notes that the Commission's report found that the operating costs of mandatory country of origin labelling for the meats under its remit would be relatively minor;

Processed meat

14. Highlights that the Commission's report of 17 December 2013 regarding the mandatory indication of the country of origin or place of provenance for meat used as an ingredient recognizes that more than 90% of consumer respondents consider it important that meat origin be labelled on processed food products;
15. Considers that consumers, like many professionals, are in favour of the mandatory labelling of meat in processed products and that such a measure would make it possible to maintain consumer confidence in food products by introducing greater transparency into the supply chain;
16. Emphasizes that it is in the interest of the European consumer to have mandatory origin labelling on all food products;
17. Points out that labelling in itself does not provide a safeguard against fraud, and highlights the need for a cost efficient control system in order to ensure consumer trust;
18. Recalls that voluntary labelling schemes, where appropriately implemented in various Member States, have been successful for both consumer information and for producers;
19. Is of the view that the failure to adopt implementing acts pursuant to Article 26(3) of Regulation (EU) 1169/2011 means that the Article cannot be properly enforced;
20. Notes that protected designations of origin already exist for many processed meat and dairy products (e.g. ham and cheese), according to which the origin of the meat used is laid down in the production criteria and increased traceability applies; calls, therefore, on the Commission to promote the development of products with "protected designation of origin" (PDO), "protected geographical indication" (PGI), or "traditional specialty guaranteed" (TSG) pursuant to Regulation (EU) 1151/2012 (7) and thereby to ensure

that consumers have access to high-quality products of safe provenance;

21. Calls on the Commission to ensure that any current EU country-of-origin labelling Regulations are not weakened in any ongoing trade negotiations such as TTIP, and that the right to propose further additional country-of-origin labelling Regulations in the future for other food products is not impeded;

Conclusions

22. Calls on the Commission to implement the mandatory indication of country of origin or place of provenance for all kinds of drinking milk, dairy products and meat products, and to consider extending the mandatory indication of country of origin or place of provenance to other single-ingredient foods or those with one main ingredient, by making legislative proposals in these areas;

23. Urges the Commission to submit legislative proposals making the indication of the origin of meat in processed foods mandatory in order to ensure greater transparency throughout the food chain and to better inform European consumers in the wake of the horsemeat scandal and other cases of food fraud; points out in, addition, that mandatory labelling requirements should take into account the principle of proportionality and the administrative burden for food business operators and enforcement authorities;

24. Considers that the aim of mandatory food origin labelling is to restore consumer confidence in food products; calls on the Commission to make a proposal to this end while taking into account the transparency of the information and its legibility for consumers, the economic viability of European businesses and the purchasing power of consumers;

25. Highlights the importance of a level playing field on the internal market and implores the Commission to take this into account when discussing rules regarding mandatory origin labelling;

26. Calls on the Commission to support labelling schemes relating to animal welfare during cultivation, transport and slaughter;

27. Deplores the fact that the Commission has still not made any move to include eggs and egg products in the list of foods for which indication of the country of origin or place of provenance is mandatory, even though cheap egg products made from liquid or dried eggs which are primarily used in processed foods

are being imported into the EU market from third countries and are clearly circumventing the EU ban on cage rearing; takes the view, therefore, that in this context the mandatory labelling of egg products and foods containing eggs to indicate origin and rearing method could improve transparency and protection, and calls on the Commission to submit a market analysis and, if necessary, to draw up appropriate legislative proposals;

28. Believes that country of origin labelling for drinking milk, lightly processed dairy products (such as cheese and cream) and lightly processed meat products (such as bacon and sausages) would have significantly reduced associated costs, and that this labelling should be explored as a priority;

29. Considers that origin labelling as such does not prevent fraud; advocates, in this connection, that a resolute course should be taken to step up monitoring, improve enforcement of existing legislation and impose more stringent penalties;

30. Calls on the Commission to take the necessary action to combat fraud in relation to rules on the voluntary labelling of origin for foodstuffs;

31. Invites the Commission to support the existing quality schemes for agricultural products and foodstuffs covered by Regulation (EU) 1151/2012, and asks for European promotion campaigns on those products to be stepped up;

32. Reiterates its call on the Commission to fulfil its legal obligation to adopt, by 13 December 2013, the implementing acts necessary for the proper enforcement of Article 26(3) of Regulation (EU) 1169/2011, so that the national authorities can impose the relevant penalties.

APPENDIX II – NUTRITION CLAIMS AND CONDITIONS APPLYING TO THEM (ANNEX TO THE REGULATION (EU) 1924/2006)

ENERGY

A claim that a food is low in energy, and any claim likely to have the same meaning for the consumer, may only be made where the product does not contain more than 40 kcal (170 kJ)/100 g for solids or more than 20 kcal (80 kJ)/100 ml for liquids. For table-top sweeteners the limit of 4 kcal (17 kJ)/portion, with equivalent sweetening properties to 6 g of sucrose (approximately 1 teaspoon of sucrose), applies.

ENERGY-REDUCED

A claim that a food is energy-reduced, and any claim likely to have the same meaning for the consumer, may only be made where the energy value is reduced by at least 30%, with an indication of the characteristic(s) which make(s) the food reduced in its total energy value.

ENERGY-FREE

A claim that a food is energy-free, and any claim likely to have the same meaning for the consumer, may only be made where the product does not contain more than 4 kcal (17 kJ)/100 ml. For table-top sweeteners the limit of 0.4 kcal (1.7 kJ)/portion, with equivalent sweetening properties to 6 g of sucrose (approximately 1 teaspoon of sucrose), applies.

LOW FAT

A claim that a food is low in fat, and any claim likely to have the same meaning for the consumer, may only be made where the product contains no more than 3 g of fat per 100 g for solids or 1.5 g of fat per 100 ml for liquids (1.8 g of fat per 100 ml for semi-skimmed milk).

FAT-FREE

A claim that a food is fat-free, and any claim likely to have the same meaning for the consumer, may only be made where the product contains no more than 0.5 g of fat per 100 g or 100 ml. However, claims expressed as "X% fat-free" shall be prohibited.

LOW SATURATED FAT

A claim that a food is low in saturated fat, and any claim likely to have the same meaning for the consumer, may only be made if the sum of saturated fatty acids and trans-fatty acids in the product does not exceed 1.5 g per 100 g for solids or 0.75 g/100 ml for liquids and in either case the sum of saturated fatty acids and trans-fatty acids must not provide more than 10% of energy.

SATURATED FAT-FREE

A claim that a food does not contain saturated fat, and any claim likely to have the same meaning for the consumer, may only be made where the sum of saturated fat and trans-fatty acids does not exceed 0.1 g of saturated fat per 100 g or 100 ml.

LOW SUGARS

A claim that a food is low in sugars, and any claim likely to have the same meaning for the consumer, may only be made where the product contains no more than 5 g of sugars per 100 g for solids or 2.5 g of sugars per 100 ml for liquids.

SUGARS-FREE

A claim that a food is sugars-free, and any claim likely to have the same meaning for the consumer, may only be made where the product contains no more than 0.5 g of sugars per 100 g or 100 ml.

WITH NO ADDED SUGARS

A claim stating that sugars have not been added to a food, and any claim likely to have the same meaning for the consumer, may only be made where the product does not contain any added mono- or disaccharides or any other food used for its sweetening properties. If sugars are naturally present in the food, the following indication should also appear on the label: "CONTAINS NATURALLY OCCURRING SUGARS".

LOW SODIUM/SALT

A claim that a food is low in sodium/salt, and any claim likely to have the same meaning for the consumer, may only be made where the product contains no more than 0.12 g of sodium, or the equivalent value for salt, per 100 g or per 100 ml. For waters, other than natural mineral waters falling within the scope of Directive 80/777/EEC, this value should not exceed 2 mg of sodium per 100 ml.

VERY LOW SODIUM/SALT

A claim that a food is very low in sodium/salt, and any claim likely to have the same meaning for the consumer, may only be made where the product contains no more than 0.04 g of sodium, or the equivalent value for salt, per 100 g or per 100 ml. This claim shall not be used for natural mineral waters and other waters.

SODIUM-FREE or SALT-FREE

A claim that a food is sodium-free or salt-free, and any claim likely to have the same meaning for the consumer, may only be made where the product contains no more than 0.005 g of sodium, or the equivalent value for salt, per 100 g.

SOURCE OF FIBRE

A claim that a food is a source of fibre, and any claim likely to have the same meaning for the consumer, may only be made where the product contains at least 3 g of fibre per 100 g or at least 1.5 g of fibre per 100 kcal.

HIGH FIBRE

A claim that a food is high in fibre, and any claim likely to have the same meaning for the consumer, may only be made where the product contains at least 6 g of fibre per 100 g or at least 3 g of fibre per 100 kcal.

SOURCE OF PROTEIN

A claim that a food is a source of protein, and any claim likely to have the same meaning for the consumer, may only be made where at least 12% of the energy value of the food is provided by protein.

HIGH PROTEIN

A claim that a food is high in protein, and any claim likely to have the same meaning for the consumer, may only be made where at least 20% of the energy value of the food is provided by protein.

SOURCE OF [NAME OF VITAMIN/S] AND/ OR [NAME OF MINERAL/S]

A claim that a food is a source of vitamins and/or minerals, and any claim likely to have the same meaning for the consumer, may only be made where the product contains at least a significant amount as defined in the Annex to Directive 90/496/EEC or an amount provided for by derogations granted according to Article 6 of Regulation (EU) No 1925/2006 of the European Parliament and of the Council of 20 December 2006 on the addition of vitamins and minerals and of certain other substances to foods (O J L 404, 30.12.2006, p. 26).

HIGH [NAME OF VITAMIN/S] AND/ OR [NAME OF MINERAL/S]

A claim that a food is high in vitamins and/or minerals, and any claim likely to have the same meaning for the consumer, may only be made where the product contains at least twice the value of "source of [NAME OF VITAMIN/S] and/or [NAME OF MINERAL/S]".

CONTAINS [NAME OF THE NUTRIENT OR OTHER SUBSTANCE]

A claim that a food contains a nutrient or another substance, for which specific conditions are not laid down in this Regulation, or any claim likely to have the same meaning for the consumer, may only be made where the product complies with all the applicable provisions of this Regulation, and in particular Article 5. For vitamins and minerals the conditions of the claim "source of" shall apply.

INCREASED [NAME OF THE NUTRIENT] A claim stating that the content in one or more nutrients, other than vitamins and minerals, has been increased, and any claim likely to have the same meaning for the consumer, may only be made where the product meets the conditions for the claim "source of" and the increase in content is at least 30% compared to a similar product.

REDUCED [NAME OF THE NUTRIENT] A claim stating that the content in one or more nutrients has been reduced, and any claim likely to have the same meaning for the consumer, may only be made where the reduction in content is at least 30% compared to a similar product, except for micronutrients, where a 10% difference in the reference values as set in Directive 90/496/EEC shall be acceptable, and for sodium, or the equivalent value for salt, where a 25% difference shall be acceptable.

LIGHT/LITE A claim stating that a product is "light" or "lite", and any claim likely to have the same meaning for the consumer, shall follow the same conditions as those set for the term "reduced"; the claim shall also be accompanied by an indication of the characteristic(s) which make(s) the food "light" or "lite".

NATURALLY/NATURAL Where a food naturally meets the condition(s) laid down in this Annex for the use of a nutritional claim, the term "naturally/natural" may be used as a prefix to the claim

APPENDIX III – HYGIENE OF PRIMARY FOOD PRODUCTION (ANNEX I OF REGULATION (EC) 852/2004

Part A: General Hygiene Provisions for Primary Production and Associated Operations; Part B: Recommendations for Guide to Good Hygiene Practice).

PART A: GENERAL HYGIENE PROVISIONS FOR PRIMARY PRODUCTION AND ASSOCIATED OPERATIONS

I. Scope

1. This Annex applies to primary production and the following associated operations:
   (a) the transport, storage and handling of primary products at the place of production, provided that this does not substantially alter their nature;
   (b) the transport of live animals, where this is necessary to achieve the objectives of this Regulation; and
   (c) in the case of products of plant origin, fishery products and wild game, transport operations to deliver primary products, the nature of which has not been substantially altered, from the place of production to an establishment.

II. Hygiene provisions

2. As far as possible, food business operators are to ensure that primary products are protected against contamination, having regard to any processing that primary products will subsequently undergo.

3. Notwithstanding the general duty laid down in paragraph 2, food business operators are to comply with appropriate Community and national legislative provisions relating to the control of hazards in primary production and associated operations, including:

    (a) measures to control contamination arising from the air, soil, water, feed, fertilizers, veterinary medicinal products, plant protection products and biocides and the storage, handling and disposal of waste;
    and

    (b) measures relating to animal health and welfare and plant health that have implications for human health, including programs for the monitoring and control of zoonoses and zoonotic agents.

4. Food business operators rearing, harvesting or hunting animals or producing primary products of animal origin are to take adequate measures, as appropriate:

    (a) to keep any facilities used in connection with primary production and associated operations, including facilities used to store and handle feed, clean and, where necessary after cleaning, to disinfect them in an appropriate manner;

    (b) to keep clean and, where necessary after cleaning, to disinfect, in an appropriate manner, equipment, containers, crates, vehicles and vessels;

    (c) as far as possible to ensure the cleanliness of animals going to slaughter and, where necessary, production animals;

    (d) to use potable water, or clean water, whenever necessary to prevent contamination;

    (e) to ensure that staff handling foodstuffs are in good health and undergo training on health risks;

    (f) as far as possible to prevent animals and pests from causing contamination;

    (g) to store and handle waste and hazardous substances so as to prevent contamination;

    (h) to prevent the introduction and spread of contagious diseases transmissible to humans through food, including by taking precautionary measures when introducing new animals and reporting suspected outbreaks of such diseases to the competent authority;

(i) to take account of the results of any relevant analyses carried out on samples taken from animals or other samples that have importance to human health; and

(j) to use feed additives and veterinary medicinal products correctly, as required by the relevant legislation.

5. Food business operators producing or harvesting plant products are to take adequate measures, as appropriate:

(a) to keep clean and, where necessary after cleaning, to disinfect, in an appropriate manner, facilities, equipment, containers, crates, vehicles and vessels;

(b) to ensure, where necessary, hygienic production, transport and storage conditions for, and the cleanliness of, plant products;

(c) to use potable water, or clean water, whenever necessary to prevent contamination;

(d) to ensure that staff handling foodstuffs are in good health and undergo training on health risks;

(e) as far as possible to prevent animals and pests from causing contamination;

(f) to store and handle wastes and hazardous substances so as to prevent contamination;

(g) to take account of the results of any relevant analyses carried out on samples taken from plants or other samples that have importance to human health; and

(h) to use plant protection products and biocides correctly, as required by the relevant legislation.

6. Food business operators are to take appropriate remedial action when informed of problems identified during official controls.

III. Record-keeping

7. Food business operators are to keep and retain records relating to measures put in place to control hazards in an appropriate manner and for an appropriate period, commensurate with the nature and size of the food business. Food business operators are to make relevant information contained in these records available

to the competent authority and receiving food business operators on request.

8. Food business operators rearing animals or producing primary products of animal origin are, in particular, to keep records on:
   (a) the nature and origin of feed fed to the animals;
   (b) veterinary medicinal products or other treatments administered to the animals, dates of administration and withdrawal periods;
   (c) the occurrence of diseases that may affect the safety of products of animal origin;
   (d) the results of any analyses carried out on samples taken from animals or other samples taken for diagnostic purposes, that have importance for human health;
   and
   (e) any relevant reports on checks carried out on animals or products of animal origin.

9. Food business operators producing or harvesting plant products are, in particular, to keep records on:
   (a) any use of plant protection products and biocides;
   (b) any occurrence of pests or diseases that may affect the safety of products of plant origin;
   and
   (c) the results of any relevant analyses carried out on samples taken from plants or other samples that have importance to human health.

10. The food business operators may be assisted by other persons, such as veterinarians, agronomists and farm technicians, with the keeping of records.

PART B: RECOMMENDATIONS FOR GUIDES TO GOOD HYGIENE PRACTICE

1. National and Community guides referred to in Articles 7 to 9 of this Regulation should contain guidance on good hygiene practice for the control of hazards in primary production and associated operations.

2. Guides to good hygiene practice should include appropriate information on hazards that may arise in primary production and associated operations and actions to control hazards, including relevant measures set out in Community and national legislation or national and Community programs. Examples of such hazards and measures may include:

   (a) the control of contamination such as mycotoxins, heavy metals and radioactive material;

   (b) the use of water, organic waste and fertilisers;

   (c) the correct and appropriate use of plant protection products and biocides and their traceability;

   (d) the correct and appropriate use of veterinary medicinal products and feed additives and their traceability;

   (e) the preparation, storage, use and traceability of feed;

   (f) the proper disposal of dead animals, waste and litter;

   (g) protective measures to prevent the introduction of contagious diseases transmissible to humans through food, and any obligation to notify the competent authority;

   (h) procedures, practices and methods to ensure that food is produced, handled, packed, stored and transported under appropriate hygienic conditions, including effective cleaning and pest-control;

   (i) measures relating to the cleanliness of slaughter and production animals; and

   (j) measures relating to record-keeping.

APPENDIX IV – GENERAL HYGIENE REQUIREMENTS FOR ALL FOOD BUSINESS OPERATORS

(Except When Annex 1 Applies) (Annex 2 of EC Regulation (EC) 852/2004)

INTRODUCTION

CHAPTER I

General requirements for food premises (other than those specified in chapter iii)

1. Food premises are to be kept clean and maintained in good repair and condition.

2. The layout, design, construction, siting and size of food premises are to:

    (a) permit adequate maintenance, cleaning and/or disinfection, avoid or minimize air-borne contamination, and provide adequate working space to allow for the hygienic performance of all operations;

    (b) be such as to protect against the accumulation of dirt, contact with toxic materials, the shedding of particles into food and the formation of condensation or undesirable mould on surfaces;

(c) permit good food hygiene practices, including protection against contamination and, in particular, pest control; and

(d) where necessary, provide suitable temperature-controlled handling and storage conditions of sufficient capacity for maintaining foodstuffs at appropriate temperatures and designed to allow those temperatures to be monitored and, where necessary, recorded.

3. An adequate number of flush lavatories are to be available and connected to an effective drainage system. Lavatories are not to open directly into rooms in which food is handled.

4. An adequate number of washbasins are to be available, suitably located and designated for cleaning hands. Washbasins for cleaning hands are to be provided with hot and cold running water, materials for cleaning hands and for hygienic drying. Where necessary, the facilities for washing food are to be separate from the hand-washing facility.

5. There is to be suitable and sufficient means of natural or mechanical ventilation. Mechanical airflow from a contaminated area to a clean area is to be avoided. Ventilation systems are to be so constructed as to enable filters and other parts requiring cleaning or replacement to be readily accessible.

6. Sanitary conveniences are to have adequate natural or mechanical ventilation.

7. Food premises are to have adequate natural and/or artificial lighting.

8. Drainage facilities are to be adequate for the purpose intended. They are to be designed and constructed to avoid the risk of contamination. Where drainage channels are fully or partially open, they are to be so designed as to ensure that waste does not flow from a contaminated area towards or into a clean area, in particular an area where foods likely to present a high risk to the final consumer are handled.

9. Where necessary, adequate changing facilities for personnel are to be provided.

10. Cleaning agents and disinfectants are not to be stored in areas where food is handled.

CHAPTER II
Specific requirements in rooms where foodstuffs are prepared, treated or processed (excluding dining areas and those premises specified in chapter III)

1. In rooms where food is prepared, treated or processed (excluding dining areas and those premises specified in Chapter III, but including rooms contained in means of transport) the design and layout are to permit good food hygiene practices, including protection against contamination between and during operations. In particular:
    (a) floor surfaces are to be maintained in a sound condition and be easy to clean and, where necessary, to disinfect. This will require the use of impervious, non-absorbent, washable and non-toxic materials unless food business operators can satisfy the competent authority that other materials used are appropriate. Where appropriate, floors are to allow adequate surface drainage;
    (b) wall surfaces are to be maintained in a sound condition and be easy to clean and, where necessary, to disinfect. This will require the use of impervious, non-absorbent, washable and non-toxic materials and require a smooth surface up to a height appropriate for the operations unless food business operators can satisfy the competent authority that other materials used are appropriate;
    (c) ceilings (or, where there are no ceilings, the interior surface of the roof) and overhead fixtures are to be constructed and finished so as to prevent the accumulation of dirt and to reduce condensation, the growth of undesirable mould and the shedding of particles;
    (d) windows and other openings are to be constructed to prevent the accumulation of dirt. Those which can be opened to the outside environment are, where necessary, to be fitted with insect-proof screens which can be easily removed for cleaning. Where open windows would result in contamination, windows are to remain closed and fixed during production;
    (e) doors are to be easy to clean and, where necessary, to disinfect. This will require the use of smooth and non-absorbent surfaces unless food business operators can satisfy the competent authority that other materials used are appropriate; and

227

    (f) surfaces (including surfaces of equipment) in areas where foods are handled and in particular those in contact with food are to be maintained in a sound condition and be easy to clean and, where necessary, to disinfect. This will require the use of smooth, washable corrosion resistant and non-toxic materials, unless food business operators can satisfy the competent authority that other materials used are appropriate.

2. Adequate facilities are to be provided, where necessary, for the cleaning, disinfecting and storage of working utensils and equipment. These facilities are to be constructed of corrosion-resistant materials, be easy to clean and have an adequate supply of hot and cold water.

3. Adequate provision is to be made, where necessary, for washing food. Every sink or other such facility provided for the washing of food is to have an adequate supply of hot and/or cold potable water consistent with the requirements of Chapter VII and be kept clean and, where necessary, disinfected.

CHAPTER III

Requirements for movable and/or temporary premises (such as marquees, market stalls, mobile sales vehicles), premises used primarily as a private dwelling-house but where foods are regularly prepared for placing on the market and vending machines

1. Premises and vending machines are, so far as is reasonably practicable, to be so sited, designed, constructed and kept clean and maintained in good repair and condition as to avoid the risk of contamination, in particular by animals and pests.

2. In particular, where necessary:

    (a) appropriate facilities are to be available to maintain adequate personal hygiene (including facilities for the hygienic washing and drying of hands, hygienic sanitary arrangements and changing facilities);

    (b) surfaces in contact with food are to be in a sound condition and be easy to clean and, where necessary, to disinfect. This will require the use of smooth, washable, corrosion-resistant and non-toxic materials, unless food business operators can satisfy the competent authority that other materials used are appropriate;

(c) adequate provision is to be made for the cleaning and, where necessary, disinfecting of working utensils and equipment;

(d) where foodstuffs are cleaned as part of the food business' operations, adequate provision is to be made for this to be undertaken hygienically;

(e) an adequate supply of hot and/or cold potable water is to be available;

(f) adequate arrangements and/or facilities for the hygienic storage and disposal of hazardous and/or inedible substances and waste (whether liquid or solid) are to be available;

(g) adequate facilities and/or arrangements for maintaining and monitoring suitable food temperature conditions are to be available;

(h) foodstuffs are to be so placed as to avoid the risk of contamination so far as is reasonably practicable.

CHAPTER IV

Transport

1. Conveyances and/or containers used for transporting foodstuffs are to be kept clean and maintained in good repair and condition to protect foodstuffs from contamination and are, where necessary, to be designed and constructed to permit adequate cleaning and/or disinfection.

2. Receptacles in vehicles and/or containers are not to be used for transporting anything other than foodstuffs where this may result in contamination.

3. Where conveyances and/or containers are used for transporting anything in addition to foodstuffs or for transporting different foodstuffs at the same time, there is, where necessary, to be effective separation of products.

4. Bulk foodstuffs in liquid, granulate or powder form are to be transported in receptacles and/or containers/tankers reserved for the transport of foodstuffs. Such containers are to be marked in a clearly visible and indelible fashion, in one or more Community languages, to show that they are used for the transport of foodstuffs, or are to be marked "for foodstuffs only".

5. Where conveyances and/or containers have been used for transporting anything other than foodstuffs or for transporting

different foodstuffs, there is to be effective cleaning between loads to avoid the risk of contamination.

6. Foodstuffs in conveyances and/or containers are to be so placed and protected as to minimise the risk of contamination.

7. Where necessary, conveyances and/or containers used for transporting foodstuffs are to be capable of maintaining foodstuffs at appropriate temperatures and allow those temperatures to be monitored.

CHAPTER V

Equipment requirements

1. All articles, fittings and equipment with which food comes into contact are to:

   (a) be effectively cleaned and, where necessary, disinfected. Cleaning and disinfection are to take place at a frequency sufficient to avoid any risk of contamination;

   (b) be so constructed, be of such materials and be kept in such good order, repair and condition as to minimise any risk of contamination;

   (c) with the exception of non-returnable containers and packaging, be so constructed, be of such materials and be kept in such good order, repair and condition as to enable them to be kept clean and, where necessary, to be disinfected; and

   (d) be installed in such a manner as to allow adequate cleaning of the equipment and the surrounding area.

2. Where necessary, equipment is to be fitted with any appropriate control device to guarantee fulfilment of this Regulation's objectives.

3. Where chemical additives have to be used to prevent corrosion of equipment and containers, they are to be used in accordance with good practice.

CHAPTER VI

Food waste

1. Food waste, non-edible by-products and other refuse are to be removed from rooms where food is present as quickly as possible, so as to avoid their accumulation.

2. Food waste, non-edible by-products and other refuse are to be deposited in closable containers, unless food business operators can demonstrate to the competent authority that other types of containers or evacuation systems used are appropriate. These containers are to be of an appropriate construction, kept in sound condition, be easy to clean and, where necessary, to disinfect.
3. Adequate provision is to be made for the storage and disposal of food waste, non-edible by-products and other refuse. Refuse stores are to be designed and managed in such a way as to enable them to be kept clean and, where necessary, free of animals and pests.
4. All waste is to be eliminated in a hygienic and environmentally friendly way in accordance with Community legislation applicable to that effect, and is not to constitute a direct or indirect source of contamination.

CHAPTER VII
Water supply

(a) There is to be an adequate supply of potable water, which is to be used whenever necessary to ensure that foodstuffs are not contaminated;
(b) Clean water may be used with whole fishery products.

Clean seawater may be used with live bivalve molluscs, echinoderms, tunicates and marine gastropods; clean water may also be used for external washing.

When clean water is used, adequate facilities and procedures are to be available for its supply to ensure that such use is not a source of contamination for the foodstuff.

1. Where non-potable water is used, for example for fire control, steam production, refrigeration and other similar purposes, it is to circulate in a separate duly identified system. Non-potable water is not to connect with, or allow reflux into, potable water systems.
2. Recycled water used in processing or as an ingredient is not to present a risk of contamination. It is to be of the same standard as potable water, unless the competent authority is satisfied that the quality of the water cannot affect the wholesomeness of the foodstuff in its finished form.

3. Ice which comes into contact with food or which may contaminate food is to be made from potable water or, when used to chill whole fishery products, clean water. It is to be made, handled and stored under conditions that protect it from contamination.
4. Steam used directly in contact with food is not to contain any substance that presents a hazard to health or is likely to contaminate the food.
5. Where heat treatment is applied to foodstuffs in hermetically sealed containers it is to be ensured that water used to cool the containers after heat treatment is not a source of contamination for the foodstuff.

CHAPTER VIII

Personal hygiene

1. Every person working in a food-handling area is to maintain a high degree of personal cleanliness and is to wear suitable, clean and, where necessary, protective clothing.
2. No person suffering from, or being a carrier of a disease likely to be transmitted through food or afflicted, for example, with infected wounds, skin infections, sores or diarrhoea is to be permitted to handle food or enter any food-handling area in any capacity if there is any likelihood of direct or indirect contamination. Any person so affected and employed in a food business and who is likely to come into contact with food is to report immediately the illness or symptoms, and if possible their causes, to the food business operator.

CHAPTER IX

Provisions applicable to foodstuffs

1. A food business operator is not to accept raw materials or ingredients, other than live animals, or any other material used in processing products, if they are known to be, or might reasonably be expected to be, contaminated with parasites, pathogenic micro-organisms or toxic, decomposed or foreign substances to such an extent that, even after the food business operator had hygienically applied normal sorting and/or preparatory or processing procedures, the final product would be unfit for human consumption.

2. Raw materials and all ingredients stored in a food business are to be kept in appropriate conditions designed to prevent harmful deterioration and protect them from contamination.

3. At all stages of production, processing and distribution, food is to be protected against any contamination likely to render the food unfit for human consumption, injurious to health or contaminated in such a way that it would be unreasonable to expect it to be consumed in that state.

4. Adequate procedures are to be in place to control pests. Adequate procedures are also to be in place to prevent domestic animals from having access to places where food is prepared, handled or stored (or, where the competent authority so permits in special cases, to prevent such access from resulting in contamination).

5. Raw materials, ingredients, intermediate products and finished products likely to support the reproduction of pathogenic microorganisms or the formation of toxins are not to be kept at temperatures that might result in a risk to health. The cold chain is not to be interrupted. However, limited periods outside temperature control are permitted, to accommodate the practicalities of handling during preparation, transport, storage, display and service of food, provided that it does not result in a risk to health. Food businesses manufacturing, handling and wrapping processed foodstuffs are to have suitable rooms, large enough for the separate storage of raw materials from processed material and sufficient separate refrigerated storage.

6. Where foodstuffs are to be held or served at chilled temperatures they are to be cooled as quickly as possible following the heat-processing stage, or final preparation stage if no heat process is applied, to a temperature which does not result in a risk to health.

7. The thawing of foodstuffs is to be undertaken in such a way as to minimize the risk of growth of pathogenic microorganisms or the formation of toxins in the foods. During thawing, foods are to be subjected to run-off liquid from the thawing process may present a risk to health it is to be adequately drained. Following thawing, food is to be handled in such a manner as to minimize the risk of growth of pathogenic microorganisms or the formation of toxins.

8. Hazardous and/or inedible substances, including animal feed, are to be adequately labelled and stored in separate and secure containers.

CHAPTER X
Provisions applicable to the wrapping and packaging of foodstuffs

1. Material used for wrapping and packaging are not to be a source of contamination.
2. Wrapping materials are to be stored in such a manner that they are not exposed to a risk of contamination.
3. Wrapping and packaging operations are to be carried out so as to avoid contamination of the products. Where appropriate and in particular in the case of cans and glass jars, the integrity of the container's construction and its cleanliness is to be assured.
4. Wrapping and packaging material re-used for foodstuffs is to be easy to clean and, where necessary, to disinfect.

CHAPTER XI
Heat treatment

The following requirements apply only to food placed on the market in hermetically sealed containers:
1. any heat treatment process used to process an unprocessed product or to process further a processed product is:
   (a) to raise every party of the product treated to a given temperature for a given period of time;
      and
   (b) to prevent the product from becoming contaminated during the process;
2. to ensure that the process employed achieves the desired objectives, food business operators are to check regularly the main relevant parameters (particularly temperature, pressure, sealing and microbiology), including by the use of automatic devices;
3. the process used should conform to an internationally recognized standard (for example, pasteurization, ultra high temperature or sterilization).

CHAPTER XII
Training

Food business operators are to ensure:

1. that food handlers are supervised and instructed and/or trained in food hygiene matters commensurate with their work activity;

2. that those responsible for the development and maintenance of the procedure referred to in Article 5(1) of this Regulation or for the operation of relevant guides have received adequate training in the application of the HACCP principles; and
3. compliance with any requirements of national law concerning training programs for persons working in certain food sectors.

APPENDIX V –
LIST OF ACRONYMS

This appendix deals with the acronyms more commonly used in the present book. It is provided to facilitate the consultation by the reader of the different acronyms.

A

AAC	Administrative Assistance and Cooperation
AIM	Active and Intelligent Materials
AMFEP	Association of Manufacturers and Formulators of Enzyme Products
ARfD	Acute Reference Dose

B

BFC	Border Food/Feed Control
BPA	Bisphenol A
BSE	Bovine Spongiform Encephalopathy

C

CAS	Chemical Abstract Service
CJD	Creutzefeld–Jakob Disease

D

DRV	Dietary Reference Value

E

EC	European Commission
ECDC	European Centre for Disease Prevention and Control
EEA	European Environmental Agency
EEA	European Economic Area

EEC	Economic European Community
ECHA	European Chemicals Agency
EFSA	European Food Safety Authority
EMA	European Medicines Agency
ERA	Environmental Risk Assessment
EREN	Emerging Risk Exchange Network
ERI	Emerging Risk Identification
ERIS	Emerging Risk Identification Support
EURL	European Union Reference Laboratories
EU	European Union

F

FBO	Food Business Operator
FCM	Food Contact Material
FEIM	Food Enzymes Intake Model
FHP	Food Hygiene Package
FIR	Food Information Regulation
FOP	Front of the Pack
FSMS	Food Safety Management System
FTE	Full Time Equivalent

G

GAP	Good Agricultural Practice
GFL	General Food Law
GMO	Genetically Modified Organism

H

HACCP	Hazard Analysis and Critical Control Points

I

IPCC	International Plant Protection Convention
IUPAC	International Union of Pure and Applied Chemistry

J

JECFA	Joint FAO/WHO Expert Committee on Food Additives

L

LOD   Lowest Level of Analytical Determination

M

MoE   Margin of Exposure
MS   Member State
MRL   Maximum Residue Limit

N

NOAEL   No Adverse Effect Level
NTP   National Toxicolgical Program

O

OECD   Organisation for Economic Cooperation and Development

P

PET   Polyethylene Terephtalate
PHL   Plant Health Law
PPP   Plant Protection Product
PPPAMS   Plant Protection Products Application Management System

R

RA   Risk Assessment
RACE   Rapid Assessment of Contaminant Exposure
RARA   Risk Assessment Research Assembly
RASFF   Rapid Alert System for Food and Feed
RPA   Reference Point for Action

S

SCF   Scientific Committee for Food
SMILES   Simplified Molecular Input Line Entry System
SOP   Standard Operating Procedure
SSC   Scientific Steering Committee

T

TRACES	Trade Control and Export System
TRIS	Technical Regulation Information
THMP	Traditional Health Medicinal Product Service
TTC	Threshold of Toxicological Concern

U

UL	Upper Level
USFDA	United States Food and Drug Agency

V

vCJD	variant Creutzfeld–Jakob Disease

REFERENCES

Altieri, A., Robinson, T., Memgelers, M., Liem, D., Silano, V. and Bronzwaer, S. (2011) EFSA 15th Scientific Colloqium: Emerging risks in food-from identification to communication. *Trends in Food Science and Technology.* xx, 1–4.

Arpaia, M. M. (2017) The right to safe food: A short path to the roots of the international legal protection of food safety. *European Food and Feed Law Review.* 12(4), 335–342.

Audino, R., Silano, V. and Tigani, F. (1995) *Prodotti fitosanitari nell'agricoltura e nell'ambiente (Phyto-Sanitary Products in the Agriculture and the Environment).* Pirola. 424 pages.

Banares Vilella, S. and Vaqué, G. L. (2018) The Commission establishes the specific compositional and information requirements for *Total Diet Replacement for Weight Control Products. Commission Delegated Regulation (EU) 2017/1798. European Food and Feed Law Review.* 13(2), 108–116.

Belluzzi, G., Pallaroni, L., Paoletti, R., Poli, A., Ponghel-lini, M. and Silano, V. (2014) *Alimenti per animali nell'Unione europea. Aspetti normativi scientifici e tecnici in materia di sicurezza, efficacia ed etichettatura.* Edagricola, Milano. 352 pages.

Campos Venuti, G., Frulloni, S., Pocchiari, F., Rogani, A., Silano, V., Tabet, E. and Zapponi, G. (1985) Management of risks in the chemical and nuclear areas. *Environment International.* 10(5–6), 475–482.

Capelli, F., Silano, V. and Klaus, B. (2006) *Sicurezza alimentare nell'Unione europea (Food Safety in the European Union).* Giuffré, Milano. 599 pages.

Carbonnelle, N., Cowper, A., Lim, A., Lotta, F. and Tarr Oldfield, M. (2016) Country of origin labelling rules for prepacked food and ingredients-an international perspective. *European Food and Feed Law Review.* 11(6), 478–485.

Carreno, I. (2017) Developments on front-of-pack nutrition declaration in the EU. *European Food and Feed Law Review.* 12(4), 321–325.

Correra, C. and Silano, V. (1995) *Alimenti e bevande (Foods and Beverages).* EPC. 4000 pages.

Deluyker, H. and Silano, V. (2012) The first ten years of activity of EFSA: A success story. *EFSA Journal Special Issues.* 5–9.

Edinger, W. W. H. (2014) Food Health Law- A legal perspective on EU competence to regulate the "Healthiness" of food. *European Food and Feed Law Review.* 9(1), 11–19.

European Commission (2000) White paper on food safety and "Farm to Table" legislation action program. European Commission. Press Release Database. 12 January 2000.

European Commission (2008) Managing food safe. January 2008.

European Commission (2013) Report to the European Parliament and the Council on foods for persons suffering from carbohydrate metabolism disorders (diabetes). 11 June 2013.

European Commission (2014) Guidance document on criteria for categorization of food enzymes. 4 February 2014.

European Commission (2016a) Report to the European Parliament and the Council on young child formulae. {SWD(2016) 99 final} Brussels, 31.3.2016.

European Commission (2016b) Report to the European Parliament and Council on food intended for sports people. 15 June 2016.

European Commission (2018) The fitness check on the General Food Law (Regulation EC 178/2002).

European Commission (DG SANTE Units D1 and E1) (2018) Commission proposal for a regulation on the transparency and sustainability of the EU risk assessment in the food chain. 68th EFSA's Advisory Forum, Sofia, Bulgaria, 6 and 7 June 2018.

Finardi, C. and Derrien, C. (2016) Novel food: Where are insects (and feed…….) in Regulation (EU) 2015/2283. *European Food and Feed Law Review.* 11(2), 119–129.

Fuentes, V. R. (2017) The rapid alert system for food and feed - A critical approach. *European Food and Feed Law Review.* 12(2), 121–130.

IPSOS and London Economics Consortium (2013) Study on the functioning of voluntary food labelling schemes for consumers in the European Union EAHC/FWC/2012 86 04. December 2013.

Lahteenmaki-Uutela, A. and Grmelowà, N. (2016) European law on insects in food and feed. *European Food and Feed Law Review.* 11(1), 2–8.

Mancuso, A., Nicolandi, L., Cabedo Botella, L., Castoldi, F., Jermini, M., Wunsch, A., Bompard, C. and Pisanello, C. (2018) Analysis of flexibility principles in food safety in the EU and consequential assessment of the notification criteria (2004–2017). *European Food and Feed Review.* 3, 220–229.

Montanari, F. and Troan, D. (2017) Modernising EU policy against phytosanitary risks - The new EU plant health law. *European Food and Feed Law Review.* 12(2), 131–141.

Oltmanns, J., Licht, O., Bohlen, M.-L., Schwarz, M., Escher, S. E., Silano, V., MacLeod, M., Noteborn, J. M., Kass, G. E. N. and Merten, C. (2019) Potential emerging chemical risks in the food chain associated with substances registered under REACH. *Environmental Science: Process and Impacts.* 21 November 2019.

Paoletti, R., Poli, A., Silano, V. and Andreis, G. (2010) *Food Nutritional and Health Claims in the European Union.* Tecniche Nuove, Milano. 360 pages.

Paoletti, R., Poli, A., Silano, V. and Klaus, B. (2011a) *Nuovi alimenti ed ingredienti alimentari nell'Unione Europea.* Tecniche Nuove, Milano.

Paoletti, R., Poli, A. and Silano, V. (2011b) *Aromi alimentari nell'Unione Europa.* Tecniche Nuove, Milano.

Paoletti, R., Poli, A. and Silano, V. (2012) *Additivi alimentari nell'Unione Europea.* Chiriotti, Pinerolo. 622 pages.

Rietjens, I. M. C. M., Slob, W., Galli, C. and Silano, V. (2008) Risk assessment of botanicals and botanical preparations intended for use in food and food supplements: Emerging issues. *Toxicology Letters.* 180(2), 131–136.

Robinson, T., Silano, V., Chiusolo, A., Dorne, J., Goumperis, T., Rortais, A., Deluyker, H., Silano, V. and Liem, D. (2012) EFSA's approach to identifying emerging risks in food and feed: Taking stock and looking forward. *EFSA Journal Special Issues*. 117–123.

Salman, M., Silano, V., Heim, D. and Kreysa, J. (2012) Geographical BSE risk assessment and its impact on disease detection and dissemination. *Preventive Veterinary Medicine*. 105(4), 253–264.

Schmidt, H. (2019) Regulation EU 2018/848-the new EU organic food law. *European Food and Law Review*. 14(1), 15–28.

Silano, V. and Comba, P. (1989) Chemicals accidents: Long-term health issues. In: *Methods for Assessing and Reducing Injury from Chemical Accidents*. Bourdeau, P. and Green, G., eds. John Wiley & Sons. Ltd. pp. 211–222.

Silano, V. (1994) *Igiene e sicurezza alimentare nel mercato unico CEE (Food Hygiene and Safety in the EEC Single Market)*. Pirola. 305 pages.

Silano, M. and Silano, V. (1997) *HACCP e Controllo ufficiale dei prodotti alimentari (HACCP and Official Control of Foods Products)*. Pitagora, Bologna. 388 pages.

Silano, V. (1997) Food contamination. In: *Environment and Health in Italy*. Pensiero Scientifico, pp. 233–261.

Silano, M. and Silano, V. (1999) *Igiene degli alimenti e principi di nutrizione (Food Hygiene and Principles of Nutrition)*. Pitagora, Bologna, 512 pages.

Silano, V. (2003) Overview of the approach for the geographical risk assessment of BSE in bovine and sheep. In: *Overview of the BSE Risk Assessment of the European Commission's Scientific Steering Committee and Its TSE/BSE Ad Hoc Group Adopted Between September 1997 and April 2003*. P. Vossen, Kreysa, J. and Gall, M. eds.

Silano, V. and Silano, M. (2006) *Prodotti e Preparazioni Botaniche (Botanicals and Botanicals Preparations)*. Tecniche Nuove, Milano, 438 pages.

Silano, M. and Silano, V. (2008) The fifth anniversary of the European Food Safety Authority (EFSA) : Mission, organization, functioning and main results. *Fitoterapia*. 79(3).

Silano, V., et al. (2009) Eugloreh (2009) the status of health in the European Union. www.eugloreh.it.

Silano, V., Coppens, D. Larranaga-Guetaria, Minghetti, P. and Erhang, R. R. (2011) Regulations applicable to plant food supplements and related products in the European Union. Food & function supplements and related products in the European Union. *Food and Function*. 12/2(12), 710–719.

Silano, V., Knaap, A., Lovell, D. and Liem, D. (2012) Main achievements and challenges of the EFSA Scientific Committee since its inception. *EFSA Journal Special Issues*. 10–18.

Silano, V., Poli, A., Paoletti, R. and Dongo, D. (2013) *Etichettatura e presentazione di alimenti e bevande nell'Unione Europea, Aspetti normativi, scientifici e tecnici*. Chiriotti Editore, Pinerolo. 360 pages.

Silano, V. (2014a) *Indicazioni Nutrizionali e sulla Salute e descrittori generici in materia di alimenti e bevande*, pp. 201.

Silano, V. (2014b) EFSA science strategy: Taking stock and looking ahead. In: *Foundations of EU Food Law and Policy*. Alberto Alemanno and Simone Gabbi eds. Ashgate, pp. 23–29.

Silano, V. and Rossi, L. (2015) Safety evaluation in the European Union of flavorings, contact materials, enzymes, and processing aids in food and its evolution over time. *European Food and Feed Law*. 110(6), 402–432. ISSN 1862-2720.

Silano, M. and Silano, V. (2015) Food and feed chemical contaminants in the European Union: Regulatory, scientific and technical issues concerning chemical contaminants occurrence, risk assessment and risk management in the European Union. *Critical Reviews in Food Science and Nutrition*.

Silano, V. and Fiorani, M. (2016) *Integratori Alimentari nell'Unione Europea*. FederSalus and Tecniche Nuove. 192 pages.

Silano, V. (2017) *Aromi alimentari: valutazione della sicurezza, normativa di riferimento, modalità di indicazione in etichetta*. Editore e-book. Cibaria. 51 pages.

Silano, V. and Carnassale, M. (2018) *Nuovi Alimenti nell'Unione Europea*. FederSalus and Tecniche Nuove. 188 pages.

Spweijers, G., Bottex, B., Dusemund, B., Lugasi, A., Toth, J., Amberg-Muller, J., Galli, C., Silano, C. and Rietiens, I. C. M. (2009) Safety assessment of botanicals and botanical preparations used as ingredients in food supplements: Testing an European Food Safety Authority-tiered approach. *Molecular Nutrition and Food Research*. 53, 1–11.

Standing Committee on the Food Chain and Animal Health (2007) Guidance on the implementation of regulation n° 1924/2006 on nutrition and health claims made on foods. 14 December 2007.

TNS European Behaviour Studies Consortium (2014) Study on the impact of food information on consumers. *Decision Making*. December 2014.

Vaqué, L. G. (2014) Information to consumers on the absence or reduced presence of gluten in food. *Europan Food and Feed Law Review*. 9(6), 358–371.

Vaqué, L. G. (2015) Directive 2005/29/EC on unfair commercial practices and its application to food-related consumer protection. *Food and Feed Law Review*. 10(3), 210–221.

ANALYTICAL INDEX

I

Information systems 150, 196
Ingredients 20, 37, 38, 39, 42, 43, 44, 49,
52, 57, 60, 63, 66, 86, 92, 93, 97,
99, 121, 123, 130, 134, 141, 151,
155, 156, 179, 203, 207–210, 231,
232, 233, 244
list 49, 92
Italian Ministry of Health 80, 202
Italian system for food official
controls 80

L

Legislation implementation 20
Loyal information practices 37

M

Mandatory
alerts 45
consumer's information 35
food (label) information 37–41,
153, 210
nutrition information 46
Margin of exposure approach (MoE)
84, 167, 239
Medium acidity 70–72
Minerals 47, 48, 59, 117, 118, 119, 120,
122, 124, 125, 126, 127, 132, 133,
137, 138, 180, 201, 216, 217

N

Naturally occurring toxins 10
Net quantity declaration 44
Novel foods 20, 34, 121, 127, 128, 129,
130, 161, 181, 242
Nutritional profies 69

O

Official
food control 20, 30, 69, 70, 72, 74–79,
81, 82, 87, 110, 112, 199, 221, 243

on imported food products in the
EU market 26, 30, 31, 72–75,
82, 89, 97, 154, 211
on foods coming from the internal
EU market 76
OIE – World Organization for Animal
Health 13, 18, 144
Ordinary legisative procedure 5, 6
Organic food 143, 150, 151, 243

P

Pathogenic 232, 233
bacteria 9, 11, 12, 78, 108, 116, 143
parasites 9, 12, 108, 232
viruses 9, 12, 78, 108, 143
Persistent organic pollutants 11
Place of provenance 37, 39, 122, 143,
151, 152, 153, 154, 155, 156, 207,
209, 210
Plant protection products (PPPs) 34,
70, 76, 100, 101, 102, 103, 161,
175, 176, 201, 220, 221, 222,
223, 239
PPPs, *see* Plant Protection products
Precautionary principle 20, 25, 27, 151,
203, 220
Pre-submission advice procedure 188,
190, 198
Primary legislative acts 5–7
Prions 12–14
Processing aids 39, 91, 98, 99, 100,
160, 244
Protection of personal data 189, 194

Q

QUID 42, 155

R

Rapid Alert System for Food and Feed,
vedi RASFF
Rapid Assessment of Contaminants
Exposure, vedi RACE
RACE 83, 84, 170, 239